The Beekeepers Annual 2019

THE BEEKEEPERS ANNUAL
IS PUBLISHED BY
NORTHERN BEE BOOKS
MYTHOLMROYD,
WEST YORKSHIRE
PRINTED BY
LIGHTNING SOURCE, UK
ISBN 978-1-912271-31-3
MMXVIII

SET IN HELVETICA LT BY DM Design and Print
Cover: A worker honeybee emerging from its cell. (John Phipps)

EDITOR, JOHN PHIPPS
NEOCHORI, 24022 AGIOS NIKOLAOS,
MESSINIAS, GREECE
EMAIL manifest@runbox.com

The Beekeepers Annual 2019

Vita Bee Health

The World's Honeybee Health Experts

- **Beetle Blaster** — Small Hive Beetle Control
- **EFB&AFB diagnostic kits** — Diagnostic Kits
- **SWARM** Attractant Wipe — Attractant Wipes
- **APISTAN** / **APIGUARD** / **BEE GYM** — Varroa Control
- **vitafeed NUTRI** — Stimulatory Feeds
- **B401** — Wax Moth Treatment
- **apishield** / **apiprotect** — Hornet & Wasp Protection

After 20 years of caring for your bees
Vita (Europe) Ltd. is proud to reaffirm our commitment to honey bee health around the world and will now trade as Vita Bee Health. We look forward to many more years of keeping your bees healthy and productive.

www vitabeehealth.com

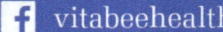 vitabeehealth

 @vitabeehealth

CONTENTS

EDITOR'S FOREWORD
THE ANNUAL: PAST AND PRESENT, *EDITOR* 3

HORNETS
INFORMATION NEEDED ON THE EUROPEAN HORNET, *Vespa crabro*, DR MICHAEL ARCHER.
NNSS POSTER: ASIAN HORNET, *Vespa velutina*, DEFRA.
NEW INCURSIONS OF ASIAN HORNET INTO THE UK, *EDITOR* 7

COLOSS
WINTER LOSSES 2016 - 2017. 11

CELEBRATIONS
WORLD BEE DAY, *RON MIKSHA, CANADA.*
TO BEE OR NOT TO BEE, *DARREN LAWS, UK.* 13

THE AFRICANISED HONEY BEES
THE KILLER BEES OF BRAZIL, *RON BROWN UK.* 19
THE MAN WHO MADE KILLER BEES, *RON MIKSHA, CANADA.* 23
THE AFRICAN BEE IN THE US, *ANN HARMAN, USA.* 28

SWARMS AND OTHER PROBLEMS
ANSWERING CALLS FOR HELP, *BILL CLARK, UK.* 32

RECORD KEEPING
THE NEXT 15 YEARS, *BILL BIELBY, UK.* 39
WE NEED TO KEEP MORE RECORDS! *JOSS LANGFORD, UK.* 40

NATIVE BEES
HISTORY OF THE NATIVE HONEY BEE IN SCOTLAND, *JOHN DURKACZ, SCOTLAND*............45
VISUAL ASSESSMENTS TO IDENTIFY NATIVE DARK HONEY BEES, *Apis mellifera mellifera*,
GAVIN RAMSAY, CHAIRMAN & JOHN DURKACZ SCOTTISH NATIVE HONEY BEE SOCIETY 51

WORKSHOP
BUILD THE ORIGINAL LANGSTROTH HIVE, *PETER SIELING, USA*..............................54

APITHERAPY
LONG FORGOTTEN REMEDIES FOR NATURAL HEALING, *JÖRG RUTHER, GERMANY.*..............62

HONEY
THE NEVER ENDING QUESTION - HOW MUCH SHOULD I SELL IT FOR? *R RAFF* 78

PUZZLES
HOW CAN 2 =1? 82
TWO WORD SEARCHES BASED ON AUTHORS OF BOOKS, PRE 1982 AND POST 1982. 83

BOOKS
NEW BOOKS PUBLISHED BY NBB IN 2018. 85

For all your Beekeeping needs...

THORNE

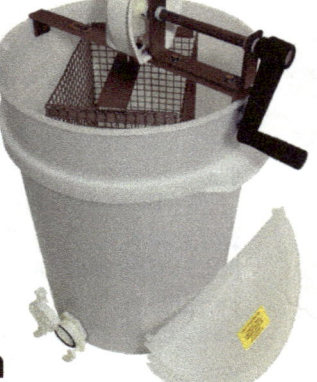

made in Britain

E H THORNE (*Beehives*) LTD
BEEHIVE BUSINESS PARK, RAND, NR. WRAGBY, LINCOLNSHIRE, LN8 5NJ
Tel. 01673 858555   sales@thorne.co.uk   www.thorne.co.uk
like us on Facebook www.facebook.com/E.H.Thorne or follow us on Twitter @ThorneBeehive

FOREWORD

John Phipps

September 2018

The Beekeepers Annual, Past and Present.

The Editor with some very gentle bees in Ukraine.

It seems such a long time ago that the first edition of The Beekeepers' Annual was published. Back in 1982, the beekeeping world was much more kindly for bees and the beekeepers were less beset by troubles compared with the many difficulties of today. Varroa, though present in Germany, had not reared its ugly head in the UK, France or Belgium; there were no Asian Hornets to worry about; the main debate about GM crops was still in the future; and problems with Colony Collapse and neonicotinoids were decades into the future. Beekeeping meetings still had the long-standing topics up for debate - should frames in the hive be placed 'warm or cold way' and how much ventilation should a hive have during winter - or was heavy insulation preferable for colony survival? There were some difficulties, of course, and these are referred to in the 1983 Annual.

The first edition of The Beekeepers Annual, 1983.

Spray Incidents

At a local level, in arable areas, spraying of oil seed rape was of great concern and Eric Fenner gave detailed information on how to send reports of spray incidents, together with samples of dead bees, either to himself or to the National Bee Unit. As time passed, local beekeeping spray liaison officers, working together with farmers and beekeepers, helped to minimise this problem as did, more positively, the ban on spraying crops in full flower.

The EEC

Whilst the UK joined the EEC in 1973, years of negotiations took place to get agreements on issues which would affect beekeeping and these were summarised by Michael Coward, who was then the General Secretary of the BBKA. By 1982, MAFF had won a concession for honey to be sold in imperial weights (though the metric equivalents had to be displayed) - which meant that no change would be needed in the jars currently used by beekeepers. As regards honey, moisture and HMF levels were set, as well as a full description of its constituents, its cleanliness and absence of fermentation. Of particular interest to beekeepers was that the EEC would grant the UK 1.2 euros for each hive in production. However, whilst in some countries this money went directly to beekeepers, in the case of the UK it went to the main beekeeping associations, pro rata, with the BBKA using the money for the Headquarters Development Fund and the erection of a building at the National Agricultural Centre at Stoneleigh. A subsidy was given on sugar for feeding bees, but this was partly due to the community's excess stocks of sugar which needed to be reduced substantially. The amount was very small: 3p per lb! During the many years that have followed, the regulations have been changed and refined involving many man-hours of work during consultations, with further regulation on bee health issues including the use of antibiotics and the introduction of bees from non-member states. What happens, one wonders after Brexit? How will all these agreements stand? Will the Secretary ultimately responsible for beekeeping give tax payers money to this important sector of agriculture? We must wait and see.

Native Bees

The importance of native bees was as important then as it is now and Ken Ibbotson, then Director of BIBBA, wrote a piece entitled "Keeping Better Bees". The brunt of his article was that the importation of bees and their hybridisation with local native bees gave rise to three very nasty factors - stinging, jumpiness (from the comb on opening a hive) and following. Whilst these types of behaviour are still common today, BIBBA have been keen to enable beekeepers to get rid of these traits with the provision of excellent record cards where these faults can be recorded, resulting hopefully in the culling of those colonies that are shown to be aggressive. A short piece followed this by Bill Bielby, CBI for Yorkshire, who was horrified that keeping records was something that was hardly done by beekeepers, so stock improvement could at best be only guesswork. Following up on this topic, thirty-six years later in this edition of The Annual, Joss Langford reinforces the importance of record keeping whether conventional or 'natural' methods of husbandry are used.

In this edition of the Annual, two pieces by John Durkacz and Gavin Ramsey of the recently formed Scottish Native Honey Bee Society, describe the society's aims and history, plus the ways in which beekeepers can learn to recognise Scottish native bee traits through visual assessments.

Hornets

Oddly enough, at a time when we are alerted to be on the look out for the Asian Hornet, a short piece in the 1983 Annual by Michael E Archer, who was based then at the University of York, undoubtedly the most important and influential entomologist of the time, was asking for samples of the European Hornet, *Vespa Crabro* to be sent to him. Part of his article is reproduced here for it is important that the range, the distribution of the native hornet, Vespa crabro, continues to be recorded (through the Bees, Wasps, Ants, Recording Society www.bwars.com) throughout the UK, so that in the event of the Asian Hornet, *Vespa velutina*, becoming established here, not only can its range be mapped but also its effect on our local hornet can be realised.

African Hybridised Bees

Undoubtedly one of the biggest stories of the time was that of the introduction of African bees into Brazil, the accidental escape of the drones, and the bees' eventual hybridisation with European bees and the rise of Killer Bees which terrified the country and spread northwards into Central America and then into the USA itself. There are three parts to this story here. The original piece by Ron Brown, who visited Brazil and who had already had knowledge of African bees during his long years in the continent; a piece more recently from Ron Miksha, who looks back on the life and work of the much maligned Warwick Estevam Kerr, the scientist behind the experiment which seemed to go so badly wrong, and Ann Harman, who looks at the current status of the African Hybridised Bee in the USA today. Whether or not beekeepers have learned to live with the difficulties this strain of bee has caused is debatable, but certainly, there seem to be two positive effects of the introduction of the bees - very high yields of honey in Brazil and the AHB's resistance to the varroa mite.

Changes to the Annual

In the past almost a third of the pages in the Annual have been taken up by the Directory with full details of the many organisations and associations concerned with beekeeping in the UK. Back in 1982, the telephone, fax, and letters were the main means of communication. Now, with almost everyone having access to the internet, information is readily available at one's finger tips. Also, during the lifetime of each Annual Directory information can get out of date very quickly. For this reason we are limiting information to the names of the organisations and associations, together with their website, so that up-to-date information can be instantly sought.

John Phipps
August 2018

HORNETS

In the 1983 Beekeepers Annual, Dr Michael Archer sent out a request for samples of the European Hornet, *Vespa crabro*, to be sent to him at the University of York, so that their distribution within the UK could be mapped.

He wrote:

In England and Wales the hornet seems to have a southern distribution reaching as far north as Shropshire. south Staffordshire. south Derbyshire. Cambridgeshire and Norfolk. Occasionally specimens are taken as far north as Yorkshire. Within the south of England it is noticeably absent or rare in Kent and East Sussex and seems to have largely disappeared from an area enclosed by Oxfordshire. north-west Essex. Cambridgeshire and Northarnptonshlre where it was formerly found. It has not been recorded from the Scilly Isles or the Channel Islands.
Typically the hornet nests in a hollow dead tree - nests have been found in willow, elm, oak, birch, beech and holly. Aerial nests have also been found in the roofs of thatched cottages and old straw stacks: in barns, garages, attics and disused huts: in bird boxes, abandoned bee hives and holes in walls. Sometimes underground nests are found. Information about the life-history of the hornet is rather limited. After the fertilised queens emerge in the spring, each queen starts to build its nest, probably at the end of May to the beginning of June. The queen rears the first workers which probably emerge in early July. The workers enlarge the nest which grows rapidly in size so that up to 200 workers will be present in a colony during August. The large cells to rear the queens are started in early August with the new queens emerging from mid-September. The drones or males also emerge at this time and the mating flights probably occur in early October. After the queens and males have left the colony the colony declines and is normally dead by late October. Probably an average-size colony rears over 300 queens."

BWARS

Information on this species **is still much in demand** and any sightings can be sent directly to the Bee, Wasps and Ants Recording Societies website. Information being sought is:
1. What is its current range and distribution?
2. Is there any sign of range change since the BWARS atlas was published?
3. What nesting sites are being used?
4. When do the first workers appear?

If you have any records of this large, colourful and distinctive social wasp, please submit the full details via this online recording platform. There is a facility for uploading photographs to support records if you have them. Please include any information on numbers, sexes, flower visitation, nesting sites and behaviour in the "Comments" section.

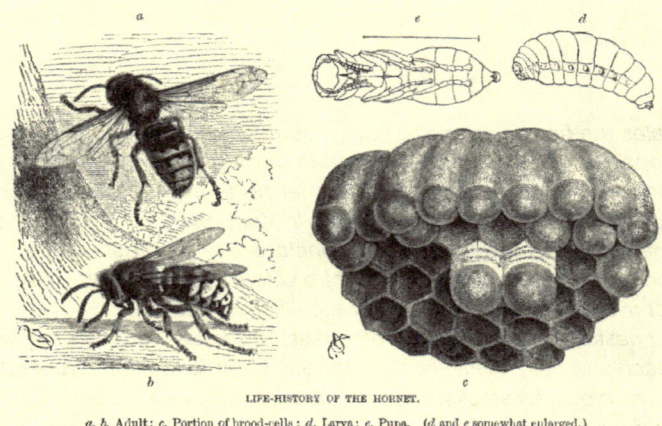

LIFE-HISTORY OF THE HORNET.
a, b, Adult; *c*, Portion of brood-cells; *d*, Larva; *e*, Pupa. (*d* and *e* somewhat enlarged.)

Left: Life History of the Hornet. (Wikipedia)

From Bees, Wasps, Ants & Allied Insects of the British Isles, by Edward Step, 1932.

It is imperative that people become aware of the differences between the European Hornet, Vespa crabro and the Asian Hornet, Vespa velutina, for if established in the UK the latter will become a major pest of honeybee colonies.

The Asian Hornet: Threats, Biology & Expansion 30 Aug 2017, jointly published by IBRA and NBB, is the most recent and authoritative work on hornets and wasps. The author, Stephen John Martin has had considerably experience with many of the species, including the Asian Hornet in its home territory.

DEFRA Poster - Opposite page

Have you seen this insect?

ASIAN HORNET
Vespa velutina

What is it?
An invasive non-native hornet originally from Asia. Suspected records should be reported immediately. A highly aggressive predator of native insects, posing a significant threat to honey bees and other pollinators. Accidentally introduced to France in 2004 where it spread rapidly. In 2016 the first UK sighting was confirmed in Gloucestershire, and a second sighting was confirmed more recently (2017) in North Devon.

Workers up to 25mm in length
Entirely dark thorax
Dark abdomen
Yellow tips to legs

Where might I see it?
Most likely to be seen close to bee hives - bee keepers should be alert. Active from February to November in suburban areas in the south of England and Wales, or around major ports.

What does it look like?
Distinctive hornet, smaller than our native species. A key feature is the almost entirely dark abdomen, except for the 4th segment which is yellow.

- Slightly smaller than native hornet
- Dark abdomen, 4th segment yellow
- Bright yellow tips to legs (native hornet dark)
- Entirely brown or black thorax (native hornet more orange)

Asian Hornet abdomen   Native Hornet abdomen

Makes very large nests

'Hawks' outside honey bee colonies killing bees as they attempt to defend their hive

DANGER!
This hornet stings. Do not disturb an active nest. Seek advice using the details below.

For more information or to report any sightings please go to:

www.nonnativespecies.org/alerts/asianhornet

or email: **alertnonnative@ceh.ac.uk**

If you have an iPhone or Android, download the free recording app:

Asian Hornet Watch

www.nonnativespecies.org

Asian Hornet Latest

Further incursions by Asian Hornets has been made into the UK. Confirmed sightings of *Vespa velutina* have been made in the Fowey area of South Cornwall, Liskaerd and Hull. Work is already underway to identify any nests, which includes deploying bee inspectors to visit local beekeepers and setting up monitoring traps.

Previous outbreaks of the Asian hornet have been successfully contained by APHA bee inspectors who promptly tracked down and destroyed the nests. The intention is to do the same in this instance.

Hopefully, nests will be found before they mature, with queens leaving the nest for hibernation quarters, ready to initiate nests next spring.

Whilst it has been in France since 2004, it reached both Guernsey and Jersey in 2016. On one occasion On one occasion in Jersey, a drone was fitted with infra red and high resolution cameras by the fire service to investigate a nest. The sound of the propellors caused the hornets to swarm out of the nest, apparently attacking the drone and spraying it with venom. This particular nest was a 'secondary' one with an estimated population of about 6000 hornets.

In Guernsey a nest of Asian hornets the size of a "small football" was found in a garden shed in L'Islet. Its location was 250m and 750m from where the hornets were sighted earlier in the year. The State of Guernsey said that efforts will be made to remove the nest as well as to locate others primarily in the St Andrew's and Longfrie areas.

COLOSS Report Winter Losses 2016 - 2017

Losses of honey bee colonies continue to cause concern in many countries. In a new paper published in the Journal of Apicultural Research, Robert Brodschneider of the University of Graz, Austria, and colleagues, present comparable loss rates of honey bee colonies during the winter of 2016-17 from 27 European countries plus Algeria, Israel and Mexico, obtained with the standardised questionnaire produced by the COLOSS association.

The 14,813 beekeepers providing valid loss data collectively wintered 425,762 colonies, and reported 21,887 colonies with unsolvable queen problems and 60,227 dead colonies after winter. Additionally, beekeepers were asked to notify of colonies lost due to natural disaster, which made up another 6,903 colonies. This resulted in an overall loss rate of 20.9% of honey bee colonies during the winter of 2016-17, with marked differences among countries.

The overall analysis showed that small operations suffered higher losses than larger ones. Overall, migratory beekeeping had no significant effect on the risk of winter loss, although there was an effect in several countries.

The paper: "Multi-country loss rates of honey bee colonies during winter 2016/2017 from the COLOSS survey" can be found at: https://www.tandfonline.com/doi/full/10.1080/00218839.2018.1460911(Open Access).

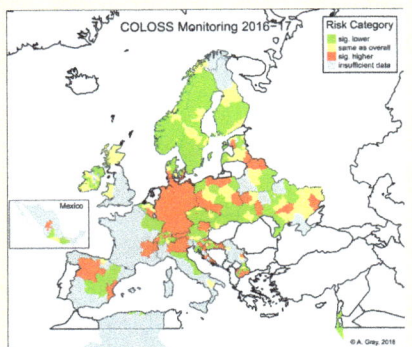

Map to show risk category in various European Countries.

POLY HIVES & POLY NUCS

Our poly hives & nucs, with range of accessories make managing swarms and colonies easy

Highly insulated nucs and full hives that help bees expand faster and overwinter better.

Durable but lightweight, making inspections Easier for all!

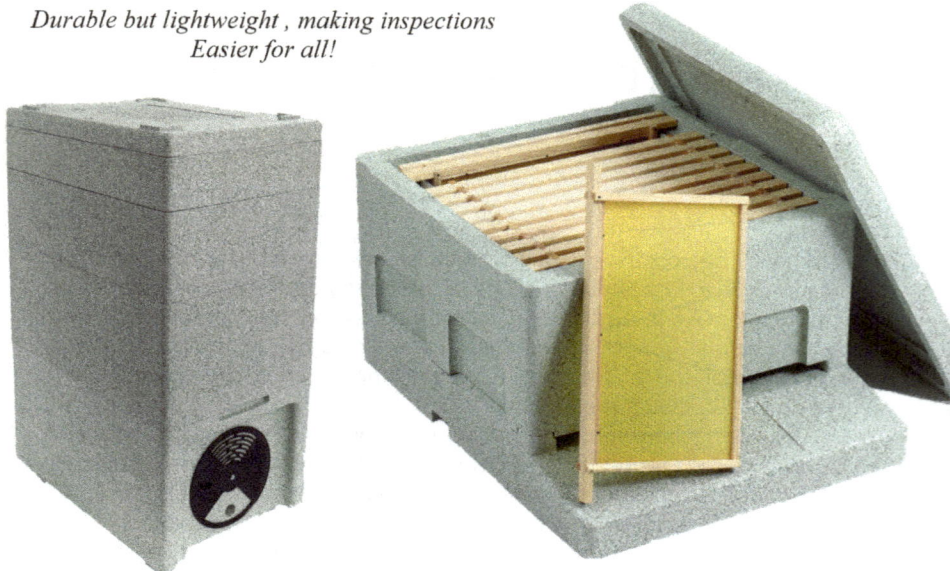

Phone or Order Online Now
OR VISIT OUR SHOP FOR THESE AND EVERYTHING NEEDED FOR YOU AND YOUR BEES

www.paynesbeefarm.co.uk
Tel. 01273 843388
Wickham Hill, Hassocks,
West Sussex, BN6 9NP

TOOLS - EQUIPMENT - CLOTHING - BEES - QUEENS - HONEY - BEESWAX

Celebrations

2018 - A Memorable Year for Bees and Beekeeping!
Time to party like it's 1734 all over again!

Ron Miksha, Canada

From this year onwards May 20th should be a big date on the beekeeper's social calendar. It's World Bee Day. Why did I mention 1734? That's the year Anton Janša was born. He was baptised on May 20th, the closest date we know to his actual birthdate. Some say that Janša was the first modern beekeeper.

Anton Janša

As an added bonus, May 20th is also another famous beekeeper's memorial day: Charles Dadant, the scion of the infinite Dadant and Sons progression of beekeepers was born in France in 1817. Dadant thought he'd be a revolutionary back in the day, in France, but he ended up in America. Charles Dadant, born on May 20, 1817, ended up in western Illinois where he wanted to grow grapes for wine. Lucky for us, his bee hives did much better than his vineyards.

World Bee Day was initiated in Slovenia, and has been quickly catching on around the world. For example, German Chancellor Angela Merkel concluded a major speech at the time with a rousing endorsement of World Bee Day, telling members of the Bundestag to do something good for the bees: *"I want to finish with something that some may consider insignificant but is actually very important: on May 20 is the first World Bee Day. On this day we should really think about biodiversity and do something good for the bees."*

World Bee Day came into being after a successful campaign by Slovenia (Anton Janša's birthplace) to promulgate the message, and their petition to the United Nations was accepted in December 2017. And beekeepers, it's your job to go out and spread the good word and "do something good for the bees". If you need some further inspiration, watch this World Bee Day video to fortify your resolve (YouTube: UN Celebrates First Ever World Bee Day 20 May) – it's about the first hive of honey bees kept at the United Nations in New York City.

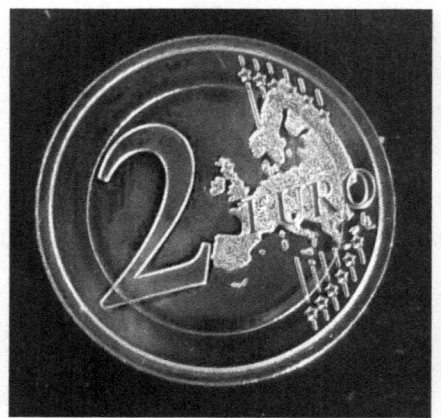 

A two-euro coin to celebrate Bee World Day, minted in Slovenia.

*For an enormous amount of background to the events leading up to and the establishment of an International World Bee Day, considerable information is given in **"No Bees, No Life"** by Peter Kozmus, Boštjan Noě and Karolina Vrtaěnik together with the 66 most important names in the field of beekeeping from 32 countries.*

It is amongst my favourites of recently published books, both in its production and content - lavishly illustrated and with up-to-date articles on a whole range of beekeeping topics. Well recommended.

Available from NBB. Editor.

To Bee or Not to Bee

28 August 2018

There really is no question about it, bees are one of the most vital components of the eco-system and in the past five years there has been a shocking decline in their population of over one third. There are many reasons for the reduction in bee numbers and together they make for worrying reading. Regardless of whether climate change is the actual cause, an increase in temperature is leading to longer periods of drought in the spring and summer. This in turn affects plants which produce fewer flowers. The fewer flowers there are the less food there is for bees.

Man has played a part in the destruction of the eco-system too by using pesticides and chemicals that have an adverse effect on insects, and the bee is not immune to the harm caused by neonicotinoids and other pesticides designed (ironically) to encourage greater crop production. The banning of neonicotinoids since 2013 on flowering plants has been a step in the right direction.

Urbanisation of the environment and city creep have also pushed the bee population back, as green spaces, wild flowers and cultivated gardens have made way for housing developments with decreasing areas set aside for nature. Against a tide of challenges, the humble bumblebee is fighting back with a little help from conservationists, bee lovers and gardeners keen to see the bee back in our gardens and once again adding to the eco-system and food chain.

The new project, 'Bee in the City' in Manchester commissioned RHS Gold award-winning garden designer, Alexandra Froggatt to take part in a celebration of our favourite insect and a symbol of Manchester's industrial revolution and more recently, a symbol that defined the spirit of the community following the tragic bombing of the Manchester arena last May. Alexandra designed a summer bee garden, commissioned by the Museum of Science and Industry, that will attract bees. Her design includes an insect hotel bench and a range of flowers designed to attract the industrious insects.

Working with Britain's leading supplier of high quality sustainable soft landscaping products, York based Rolawn provided beds and borders topsoil and soil improver, which Alex used as a mulch in the summer bee garden. *"Garden designers and gardeners everywhere need products they know will provide the best environment, especially with events like this, where the plants and flowers need to thrive quickly and look established. The garden needed to flourish quickly and all importantly, attract bees. The quality and provenance of our topsoil and mulches mean we are confident to supply leading garden designers and garden shows with products that they can trust."* Rolawn Sales and Marketing Director, Jonathan Hill said.

Alexandra Froggatt added *"The Summer bee garden is a quirky and fun outdoor space centred around a giant bee sculpture (by artist Tim Sutcliffe) which showcases simple but effective ideas the public can use to attract bees. Nectar rich plants such as Echinacea purpurea white swan, Sedum matrona, and Penstemon purple passion are mingled together to create both a beautiful display and provide food for the bees all summer and well into the autumn.*

Bees are essential in pollinating the crops we eat so a range of edibles including herbs, fruit and salads are grown to highlight this crucial connection. Many people have limited outdoor space or are on a small budget so the garden uses recycled materials in a creative way, such as turning reclaimed pallets into a living wall abundant with herbs and strawberries, proving that you don't always need a big budget to have big ideas."

from Darren.laws@zesttheagency.com

Garden Designer Alexandra Froggat. (Jason Lock)

Views of parts of the garden. NB use of pallets to create living walls. (Alexandra Froggatt)

Close-up view of one of the bee sculptures. (Jason Lock)

Ron Brown

THE AFRICANISED HONEY BEES
The Killer Bees of Brazil
Ron Brown

The Background
From time to time we read horrifying accounts and see frightening films of bees alleged to be so dangerous that they might almost threaten mankind at large. Where does the truth lie? As one might expect it is not nearly so bad as reports in the media suggest, but it is serious, at least to beekeepers. Here are the facts: honey bees were introduced into Brazil by the Portuguese in the 18th century, but were never very successful; despite a favourable environment and the vast areas available, total annual honey production was still under 5,000 tons in 1950. With the intention of breeding better bees, 28 queens were imported from Taborah (Tanzania) and Pretoria in 1957, the idea being that the climatic conditions were roughly equivalent to those in Brazil.

Accidental Queen Release
In order to prevent the escape of swarms, the hives with African queens (at Rio Claro research station in southern Brazil) had Queen Excluder strips across the entrances. However, a well-intentioned beekeeper, unaware of the experiment in progress, noticed pollen loads being lost at the congested entrance and pulled off the Queen Excluder strips. Before anyone realised what had been done, twenty-six swarms had escaped in two or three days, headed by African queens. For a year or two nothing untoward occurred, and then reports of very aggressive bees started to come in over an area of 50 or 60 square miles. By 1963 Adansonii* bees were reported in an area 500 miles across, and since then have been advancing by 150 to 250 miles a year, reaching Argentina and Guyana by 1975, Venezuela in 1976 and Panama in April 1982. At this rate they will be in Mexico before 1986 and a real threat to the USA by 1990. At first many beekeepers in Brazil gave up and stocks were burned; local authorities banned beekeeping in or near urban areas.

The Florianopolis Conference
In 1978 there was a conference on the problem in Florianopolis, Santa Catarina Province, Southern Brazil. This was attended by hundreds of beekeepers from all the S American states, plus about twenty from the USA and another dozen or so from tropical areas in other parts of the world. I was invited and, apart from one freelance journalist, was the only one from the British Isles.

No-one attended from France or Belgium, one from Portugal and Dr Koniqer from Germany; the presence of a contingent of Venezuelan fireman (in full uniform), underlined the seriousness of the situation!

For three or four days papers were read by well-known beekeeping scientists and by civil servants in the Agriculture Departments of various S American countries. Mv humble contribution (later published by Apirnondia) was a paper on "Adansonii Management" based on my twelve years beekeeping experience in Central Africa, most of the time with six to ten hives in my garden in Lusaka and Ndola, getting over 100 lb a year per hive regularly, without worrying my neighbours on either sides, and using techniques developed over several years. In these first few years, on various occasions our chickens were all killed, we were besieged by angry bees for hours at a time, and so on, before I learned by trial and error how to operate.

Alas, the Brazilians were more impressed by the American approach, dealing with very large scale operations. My plea that 'small is beautiful' and 'a cash crop for 100,000 peasants is better than 100 huge honey farms' was not taken seriously. All the same, when I spoke I noticed my audience increasing rapidly after the first minute or two until there was standing room only (loud speakers outside sent the proceedings into every room and lobby - we were using the Provincial Parliament buildings). Was this to hear the great message on how to cope with African bees? No. Although I was indeed flattered to be told why they all came in. It was, they told me later, to see who was speaking such fluent English in such a clear voice! Gratifying, in this case, to have been the only Englishman present!

The Tour
After these sessions, we visiting beekeepers were taken on a comprehensive tour of apiaries in Santa Catarina and Parana provinces, over many hundreds of miles, staying in different towns, sometimes flying, sometimes by coach. The hospitality was out of this world and the tour quite unforgettable. In general terms their approach to the problem was to site large apiaries in forest areas remote from towns, with hives on single stands (to prevent vibration exciting other stocks), using smoke on a huge scale. Their smokers were the size of small dustbins. and the volume of smoke comparable to that used to cover Infantry attack in war.

A great deal of smoke was needed to handle the hybridised bees, and colonies were best dealt with towards dusk.

I next spoke to Helmut Wiese (President of the Confederation of Brazilian Beekeepers) in Mexico last October (1981), and he told me that honey yields had greatly increased and were now up to 15,000 tons a year (Mexico and China at about 50,000 tons each are at the top of the league table). So the original purpose has been achieved, at least, but the threat to urban areas by wild colonies persists, and it is sad to realise that honeybees are pests to be destroyed when found in wild colonies in or near towns.

Why so Aggressive?

Why is the Adansonii bee so aggressive and so prone to swarm frequently and over such large distances? I think these characteristics have been programmed into them in the following manner. Africans were (and largely still are) bee hunters rather than beekeepers, and if over tens of thousands of years one kills and robs out the 'soft targets' and leaves alone the more vicious ones, then gentle bees are eliminated and those that remain are genetically 'selected' for aggressiveness.

The unusual swarming propensity can be accounted for by the shortage of reasonably-sized cavities. The trees are smaller because of frequent bush fires over most of the area in the long dry season, and I frequently found colonies of bees occupying cavities that no European bees would have been interested in, even in street lamp posts. The need for swarms to travel long distances also arises from frequent fire danger - only colonies flying twenty to thirty miles could get out of the burning area. About the distances travelled there is certainly no doubt; many times in September and October I have heard the noise of a swarm coming from a distance, like a small helicopter, flying overhead in a whirling cloud of bees at a speed of ten to fifteen mph and continuing out of sight, although on occasions I have followed for five to ten miles before losing contact. So all these factors:- frequent swarming, swarms travelling long distances, an aggressive nature, have become well-established genetically. The hostile environment (including hostile humans) has permitted only bees with these characteristics to survive.

What of the Future?
The hope that by hybridising with local bees the aggressive strain would be diminished has not so far been realised. In 1978 the bees I saw and photographed in Brazil were just like my bees in Lusaka and Ndola; slightly smaller than those in Europe, very aggressive and good honey-getters. In my opinion they present no threat to temperate climates. Nearly thirty years ago, working in Lusaka, I established their threshold working temperature at 57°F. Over many days in our Central African 'winter' I correlated a count of bee traffic at hive entrances with ambient temperature and, below 56°F, found virtually no activity. On a cloudy day in July (at 4.000ft asl), there would sometimes be a maximum temperature of less than 57°F and then scarcely a bee flew. On most days the temperature reached 57°F before midday (often much more) and the change in flight activity between 56°F and 58°F was dramatic, both in commencing and in terminating as the temperature fluctuated. At home in Torquay my bees in March behave similarly, but over a temperature range about 10° cooler (except for water-carrying, which goes on at temperatures down to 41°F or 42°F).

I have no intention of introducing Adansonii to Torquay, but if I did, their spring build-up would be so slow that they would be many weeks behind our bees and unable to compete. A threat to Mexico and the Southern States of the USA, yes. To most of North America - No!

* *Apis mellifera Adansoni or Apis mellifera Scutellata*
 - from The Beekeepers Annual, 1983.

Ron Miksha, Canada.

Warwick Estevam Kerr:
The Man Who Made Killer Bees
by Ron Miksha, Canada.

Warwick Estevam Kerr, the man who made the killer bees, celebrated his 95th birthday on September 9th, 2017. Just like his bees, Kerr comes from hot, tropical Brazil. And just like his bees, Dr Kerr has been much maligned and misunderstood in the popular press. But Kerr did more to help his country's agriculture than perhaps any other individual.

When the Africanized hybrid honey bee entered our awareness in the 1970s, the bee was described as a killer bee (in Brazil, they called it the assassin). The man who brought African honey bees to South America was turned into a mysterious fiend who had "disappeared from sight" after "he turned killer bees loose". Well, he did disappear for a while. He was in prison. But not for any reason you might guess. First, some background.

What was Kerr's crime?
Dr Warwick Kerr brought Africanized genetic stock to South America in 1956. In today's context, importing an alien creature from another continent seems horribly reckless. In Dr Kerr's day, the importation of bees from Africa was hardly daring. First, recall that all honey bees in the Americas are imported from somewhere else. Honey bees are not native to the western hemisphere. Second, Kerr was not introducing a new **species**. The African bee (*Apis mellifera scutellata*) is a cousin of a common European honey bee, *Apis mellifera iberiensis*, which was in Brazil when the African queen bees arrived. Kerr's importation of twenty-six queen bees from Tanzania is in league with importing Clydesdales long after Arabians and Morgans were already established. Kerr's goal was to improve the non-tropical honey bees which farmers were using in Brazil. He rightly assumed that tropical genetic stock would be more successful in his tropical country.

Warwick Kerr's sour reputation came directly from the Brazilian government. Although he was a geneticist and was at first entrusted with developing a better bee for Brazil's farmers, the Brazilian military dictatorship attacked Kerr's stand on civil rights. He was imprisoned in 1964 when he publicly fought government corruption. In 1969 he was re-arrested, this time for protesting that Brazilian soldiers who had raped and tortured a nun went unpunished. Sister Maurina Borges, who ran the Ribeirão Preto Orphanage, was an activist; the soldiers were part of Brazil's military dictatorship, committing crimes encouraged by the government. Most of the western press didn't bother to investigate the reasons behind the Brazilian government's dismissal of Kerr's work, his qualifications, or his imprisonment.

Creating a clown

All of this is lost on most people who write about this subject. For example, this is from a blog promoting a book called The Animal Review: A Report Card. The writer calls Dr Kerr a clown:

*"It is strange and unfortunate that there is not a **Nobel Prize for Really Bad Mistakes In Science**. This international award could be presented annually in Stockholm by a sad clown wearing a lab coat and goggles, giving scientists that much more of an incentive to get things right for once. Brazilian geneticist Warwick Estevam Kerr would have made a fine nominee. For it was Mr. Kerr who introduced Africanized honey bees (Apis mellifera scutellata) to the Americas. Oops. Bring in the clowns...The full scope of the blunder was not immediately apparent to Kerr. Being a brilliant geneticist, he brilliantly assumed the African queen fugitives would breed with feral bees — thus diluting their infamous aggression. But on the bright side, Africanised honey bees pollinate plants and plants are crucial to agriculture production everywhere in the blah, blah, blah, blah."*

"Warwick Estevam Kerr, Grade: F- "
Almost everything in the preceding story is wrong, but I put it here to illustrate how the popular press saw Dr Kerr – a clown deserving an F- grade. In fact, it's the lazy reporters who earn a big fat Fail.

Here's another example: National Geographic blunders portraying Dr Kerr in their 2006 documentary, Attack of the Killer Bee: *"Incredibly, nearly one trillion killer bees can all be traced back to just one man..."* [I'll bet you know who they're talking about.] In Africa, says NG, Kerr *"chose the best specimens he could find, but he noticed something disturbing."* (At this point, the actor playing Kerr gets stung on the finger and yelps "Ouch!" in pain. *"Doctor Kerr was wrong. Very wrong. And the western hemisphere is still paying a steep price."* This is verbiage that sells, even if utterly wrong.

You should watch the first few minutes of the following NG fantasy. The devilish portrayal of the black Africans who sold Kerr the 'deadly' bees is also vile racist National Geographic reporting, but that's fodder for a whole different story. The video "Attack of the Killer Bees" (A National Geographic Documentary) queued up to start at three minutes – shows where an actor playing Kerr gets ready to leave for Africa. But don't bother to watch more than a minute or two of this.

The Killer Bees
Warwick Kerr was responsible for bringing African genetic stock to Brazil in 1956. As a geneticist, he wanted to improve the health and hardiness of the European honey bee which came from Portugal in 1834. That European strain was poorly adapted to the tropics, so the Italian honey bee *(Apis mellifera ligustica)* was imported in the 1880s, but it wasn't much better. A few farmers and monks kept the languid bees, mostly to produce beeswax for church candles.

In 1956, Brazil's annual honey production from the European honey bees was just 15 million pounds. Brazilian agriculture was expanding and needed a tropical honey bee for pollination and honey production. After the African bees arrived, Brazil's beekeepers produced 110 million pounds. Brazil went from 43rd in the world to 7th largest honey producer. By 1994, L.A. Times headlined: *"Brazil's honey production has soared since the ornery invaders took over beekeepers' hives"*. Today, most of the world's organic honey is produced by Africanized honey bees in Brazil's remote forests. The honey is doubly organic – produced in areas untouched by pesticides and produced in Africanized hives which are naturally resistant to varroa – so mite meds aren't used in those colonies.

Honey bees with African genes are more aggressive than European bees. Beekeepers in Brazil had to learn appropriate management techniques. Although the venom is the same, more bees attack if their colony is disturbed. People have died from massive numbers of stings. Those deaths are sorrowful and this story about Dr Kerr's bees should not dishonour personal tragedies. Some of the traits which make Africanised bees exceptional pollinators (refined olfactory sense, quicker movements, flights in inclement weather, superior navigation skills) also make them more likely to sting. However, they can be managed by farmers and beekeepers. Indiscriminate killers they are not.

Decoding sex among stingless bees
At first, Warwick Kerr worked with *Melipona* bees, not honey bees. Some of Brazil's poor and indigenous were wild honey gatherers, or *meleiros*. Meleiro, isolated and rural, is named for the meleiros, who are named for *Melipona* honey trees. There are only 7,000 meleiro people, but their precarious existence in the 1940s – which included raiding *Melipona* bee trees – concerned Dr Kerr during his bee research. He hoped that his work would draw attention to the importance of preserving *Melipona*, their habitat, and the people who lived off those bees. Understand and help the Melipona, and you help the meleiros, figured Kerr.

(Melipona quadrifasciata of the meleiros (João Henrique Dittmar Filho)
Melipona quadrifasciata is a eusocial stingless bee, native to southeastern coastal Brazil. The meleiros call it Mandaçaia, which means "beautiful guard," as there are always guard bees defending the narrow entrance of their colony. Brazil's Melipona build mud hives inside hollow trees. These have narrow passages allowing just one bee to pass at a time. Stingless bees, they can give a nasty bite, but their intricate passage system also defends against predators.)

Melipona quadrifasciata. (Wikipedia, Thiago Mlaker, 2009)

Dr Kerr's first influential paper, "Genetic Determination of Castes in Melipona" (1949), researched the development of males, females, and workers among Brazil's common stingless bee. Kerr found that their caste development was different from honey bees. Drones in both species are haploid, but in *Melipona*, things get weird for the girls.

In *Apis mellifera*, "a larva develops into a queen or into a worker depending upon the food it receives. In *Melipona*, on the other hand, caste determination is genotypic. Fertile females (queens) are heterozygous in some species for two, and in other species for three pairs of genes, homozygosis for any one of which makes the individual develop into a worker." – Kerr, 1949.

For the exotic *Melipona quadrifasciata*, alleles (one-half of a gene that controls an inheritance, for example the 'b' in a 'Bb' gene) determine caste. Drones (as in honey bees) are haploids with a single set of chromosomes; queens and workers are diploid (two sets of chromosomes, one from each parent), but queens have some specific alleles that are different, or heterozygous (for example, AaBb), while workers have identical, or homozygous, caste-determining genes (AABB, AAbb, aaBB, or aabb combinations). If you find this confusing, imagine sorting it out with 1940s technology, as Kerr did.

WARWICK E. KERR

$$
\begin{aligned}
\text{Queen} \times \text{Drone} &= 1 \text{ Queen} + \quad 3 \text{ workers} \\
AaBb \times AB &= AaBb + (AABB + AABb + AaBB) \\
AaBb \times Ab &= AaBb + (AABb + AAbb + Aabb) \\
AaBb \times aB &= AaBb + (AaBB + aaBB + aaBb) \\
AaBb \times ab &= AaBb + (Aabb + aaBb + aabb)
\end{aligned}
$$

From Kerr's 1950 Melipona paper

The real Warwick Kerr

Kerr was born in São Paulo, Brazil, in 1922, into a middle-class family with Scottish roots. He received an agricultural engineering degree, then specialized in genetics. His work as an entomologist spanned decades, with research that included genetics of honey bees and native Brazilian bees, as we've just seen.

Warwick Kerr's post-doc research was at the University of California, Davis (1951), and at Columbia University in New York, under the renowned evolutionary biologist Theodosius Dobzhansky. One of Kerr's influential papers, "Experimental Studies of the Distribution of Gene Frequencies in Very Small Populations of Drosophila melanogaster", cites Dobzhansky as an adviser and is co-authored by a University of Chicago genetics statistician. This fruit fly research was done way back in 1954 and the paper was one of the first to deal with the nascent field of genetics statistics. Eventually, Kerr published 620 research papers during his 60-year career.

Warwick Kerr is largely responsible for establishing the study of genetics in Brazil. He was a director of the National Institute for Research in the Amazon and worked at the University of São Paulo. Later, at the Universidade Estadual do Maranhão, he created the Department of Biology and served as Dean of the University.

Warwick Kerr says that his most important work was developing staff, technicians, teachers, and researchers in his country. At the University of São Paulo, he established a department of genetics which focuses on entomological and human genetics, using mathematical biology and biostatistics. Kerr has memberships in the Brazilian Academy of Sciences, the Third World Academy of Science, and the US National Academy of Sciences.

The African bee in the U.S.

Ann W Harman

The African bee (*A. m. scutellata*) arrived in the United States in 1990, crossing from Mexico into the state of Texas. Unlike European stock bees (EHB), the African bee (AHB) will not only live in cavities in the ground but also in small cavities. The warm climate and the terrane of the South and Southwest provided excellent nest sites as well as sufficient forage.

This overly-defensive bee is a true survivor bee. Since the queen is a profuse egg-layer, a nest rapidly becomes overcrowded. Swarms are cast and can quickly find a new site. Nest sites acceptable to AHB are generally smaller than those chosen by EHB. However, AHB will readily nest in abandoned vehicles, discarded tires, overturned empty flower pots. They are commonly found in equipment used only seasonally. In towns and cities AHB regularly inhabits in-ground water meter cavities and hollow metal lampposts.

In the warmer climates of the US, flowering plants can be found almost year around. However, if forage becomes scarce the AHB simply absconds, leaving behind brood and even some food. The absconds, as well as swarms, can migrate, even foraging on the way, until a suitable site is found. The AHB also has another tactic to assist its spread. It is able to usurp a colony of EHB living in a beehive. A beekeeper with a hive of EHB may notice a small cluster of bees by a corner of the hive entrance. This group of bees, actually AHB with their queen hidden inside the ball of bees, is slowly invading the EHB colony. The EHB queen is killed by the invading workers or by the African queen. As soon as the original queen is dead the rest of the AHB colony simply moves in. With eggs now being laid by the African queen, the colony becomes African, much to the surprise of the beekeeper who now has an overly-defensive colony.

With its swarming, absconding and usurpation, the AHB has colonised a number of states of the US. It is also an efficient hitchhiker. Swarms frequently bivouac on vehicles. If that vehicle, with its unnoticed clump of bees, drives a long distance before stopping, a colony of AHB has now found a new place to live and colonise the surrounding area from swarms. In spite of predictions about AHB only existing in warm climates, a few years ago a colony was found in the state of Colorado, known for cold, snowy winters. This colony, identified as AHB, had successfully overwintered. It is a survivor bee.

In the US the European stock bees have to contend with imported pests; the varroa mite and the small hive beetle. EHB colonies need the beekeeper's help to survive. The AHB has evolved with the small hive beetle and does not tolerate it in the hive. Varroa has arrived in Africa but again is not causing the harm found in countries with European stock bees.

The African bee can be an excellent honey producer when given appropriate forage. AHB will also take advantage of forage with low sugar content that the EHB ignores. In the yearly lists of top honey-producing countries, Brazil and Mexico will always be found and sometimes Argentina. Their only honey bee is the African bee.

After the AHB's arrival in Brazil beekeepers who had the poorly-performing EHB simply stopped being beekeepers. The overly-defensive AHB made beekeeping difficult and unpleasant. In a short time, new beekeepers appeared. They were the ones who discovered that beekeeping practices that were usual for EHB could not be used for AHB. It is "let alone" beekeeping. Leave the colonies alone, provide them with honey supers at the appropriate time; remove the filled supers and harvest the honey. In short, do not "manage" the AHB colony as a beekeeper would do with an EHB colony. This scene—of traditional beekeepers quitting and new ones appearing—followed the AHB's progress through South America, Central America, Mexico and into the United States. Those beekeepers in the US who live in the AHB colonised areas either keep their colonies as AHB or perform rigorous re-queening with EHB queens. Re-queening an AHB colony can be quite difficult and may need to be done twice a year. The EHB queens would have to be purchased from outside the established AHB area, therefore an expense for the beekeeper.

The AHB is still advancing and colonising new areas of the US. At one time the US Department of Agriculture kept track of the advance and published a yearly map. However, that project was discontinued quite a number of years ago. Many maps found on the internet are outdated. At present the individual state apiary inspectors keep track of AHB colonisation in their particular state. States, mainly in the southern tier of states, maintain trap lines. Ports along the East and West Coasts also maintain traps.

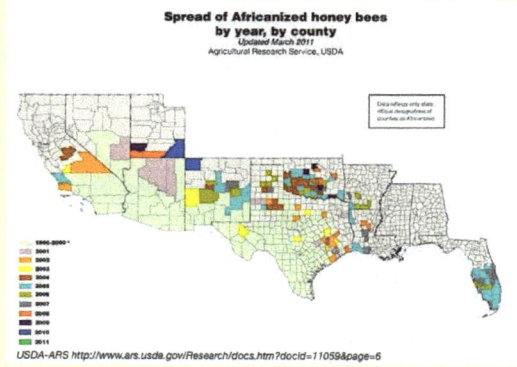

Spread of Africanized honey bees by year, by county
Updated March 2011
Agricultural Research Service, USDA

USDA-ARS http://www.ars.usda.gov/Research/docs.htm?docid=11059&page=6

Despite the great interest in the spread of AHB in the USA, there seems to be very little information on its current range.

The colonisation has been a bit puzzling over the years. AHB quickly became established throughout Texas, New Mexico, Arizona and Southern California. Today AHB is considered established throughout California except for the very northern part that probably has only feral colonies. Nevada and Utah, with hot arid conditions, are colonised. From Texas AHB moved north and colonised Oklahoma. However, Arkansas, also next to Texas, has colonisation only in a very narrow portion, just in a few southern counties. The bee finally moved into Louisiana from Texas in 2005. Both the states of Mississippi and Alabama do not have AHB except perhaps some feral colonies in the far southern tips of the states. These states maintain trap lines. However Florida has established colonisation in the entire lower half of the state with feral colonies throughout the northern half. It is thought that AHB reached Florida from the several commercial ports on the Atlantic Ocean and Gulf of Mexico coasts.

In spite of predictions of the normal range of the African honeybee, it is possible that it will be more adaptable to colder climates than we think at this time. It is also possible that in future years it will become less defensive. Puerto Rico is a small island that has had the AHB for quite a number of years. Today beekeepers find AHB to be less defensive. Perhaps the beekeepers simply helped the conversion by destroying the really over-defensive colonies. Since the genes for defensive behaviour are carried by the drones, elimination of whole colonies would remove those drones from mating.

The Latin American countries have done, and continue today, excellent public education about the AHB. The United States has done some education in the colonised areas. However incidents, such as a massive stinging of a person, will be reported by the media even outside the area concerned. These reports have become fewer over the years.

Colonies of AHB cannot be kept in urban or suburban areas. So as the bee advances, many small-scale beekeepers that exist today will have to give up their beekeeping. Commercial beekeepers will have to make some changes for honey production and pollination. These beekeepers will succeed. After all, the beekeepers in Latin American countries are able to produce honey and perform pollination services.

BeeCraft

Celebrating 1919 BeeCraft 2019 100 years

The longest running UK beekeeping publication is celebrating its centenary in 2019

See a free copy before you buy at
www.bee-craft.com/try

Choose between hard copy with complimentary digital or digital only

Sign up at www.bee-craft.com/shop
and join in our celebrations in 2019

ANSWERING CALLS FOR HELP FROM THE PUBLIC

Author Bill Clark, Cambridge

After Varroa destructor arrived in the Cambridge area, calls regarding swarms dropped by 90%, however calls about other insects increased. I wonder if folk are getting more intolerant - and fearful. Are wild bees finding the garden habitat more to their liking, and are they getting more numerous as they fill the void? Perhaps a little of all four. In most cases sensibly imparted advice is all that is needed, I am happy to say that I often persuade my caller to 'leave well alone', some have even moved on to promote insect conservation themselves!

In the first breathless moments the caller has usually imparted their vital information - whether they have been stung - but often they just ran and a vivid imagination took over. Our phone rang on a - for me - busy Easter bank holiday: 'There is an angry bees nest in our garden'. However none of my wife's proffered 'nearby' beekeepers were able to help and they next called at 5.0 pm. 'Please, please can your husband help us, we have been shut in all day, there are bees everywhere even, bouncing off the windows'! Upon my arrival, the nest's whereabouts was imparted through their letter box. A row of faces watched through a window as I approached a partly clipped bush, in the center was an old

Swarms seemed to have decreased locally since varroa became a problem.

thrush's nest; lifting some moss revealed a large red-tailed bumblebee guarding her honeypot and six brood cells - one recently vacated. I selected a shoe-box from my car, and after catching the single loudly buzzing worker, placed the lot inside. Through the letter-box I assured them there was no more danger and a ten pound note fluttered out of the slot! Not until I was driving off, did the large - shirt sleeved - father step tentatively outside!

The following list of usual descriptions, followed by possible answers, should be helpful.

Description	Possible answer
A sudden cloud of humming insects....	An incoming - or departing - honeybee swarm?
Above, litter bins, factory drums, etc...	Bees/wasps, clearing syrup, honey, sugary drinks, etc?
Coming out of the ground....................	Single 25mm - 50mm hole, wasps or bumbles?
" " " " "	Single or more 5mm - 10mm holes, solitary bees or wasps?
Only seen after being disturbed...........	Bumbles or social wasps?
Flying out of a clipped hedge...............	French-wasp nest or honeybee swarm, maybe on comb?
From a hollow tree, box, barrel, etc.....	Social wasps, hornets or honeybees nest?
Or compost, straw-bales, bird box, etc.	Either of the Red-tailed bumblebees or social wasps?
Nest under shed, rubbish, paving, etc..	Red-tailed or Carder bumblebees or social wasps?
At ground level, or in grass tussocks...	Red-tail or Carder bumblebees?
Any of above six, esp waste ground..	Buff-tailed or White-tailed bumblebees?
Flying from roof space........................…	Wasp, honey bee or hornets nest?
Hole in a wall known to be hollow......	Wasp, honeybee; four types of bumblebees? (Bumblebees love insulated cavities)
Wall ventilator bricks...................…..	Wasps, honeybees, red-tailed and now tree bumbles?
Chimney...	Honeybees, rarely wasps or hornets?
Holes in cliffs, walls and timber:	
Single, up to pencil sized ...…..........	Solitary bees/wasps? Some bees look waspish & vice versa.
Multiple as above................................	Solitary bees/wasps in a communal location?
Tween roof tiles or weather boarding..	Solitary mason bees/wasps?
Using 5 - 10 mm common entrance.....	Occasionally solitary bees - communal location inside?
20 - 55 mm common entrance.............	Social wasps or bumblebees?

Numbers issuing forth from nests can depend on weather and season. Single social wasps in May will be a constant stream in August, but even Buff-tailed bumblebees will stay inside on a cold and wet August day. Honey bees building comb in the open are no longer a rarity, and research has shown that as many as one in six social wasps nests evolve mid-season - yes from a swarm of disturbed wasps or ones finding their present cavity too small! Also be aware that non-beekeepers can describe a few dozen - mating solitary bees - hovering by their house wall or above the lawn, as, 'a great cloud'. A Red-tailed queen - B. lapidarius in a birds nest box, can sound very loud: I once observed a Great Tit - they will feed on bumblebees - rocket from a nest box in much agitation and I almost fell off the ladder when the queen buzzed as I lifted the roof off - the birds nested elsewhere! The Tree bumblebee, Bombus Hypnoram, has spread widely, and is now a very likely resident of such boxes and the Asian Hornet is another newcomer to our shores, even the beekeepers are misnaming this one!

Colour, size and shape:
A 'stream' of black and yellow, 11 - 14 mm, wasp-waisted insects, will be social wasp workers, most likely in Britain to be *Vespula vulgaris* or *V. germanica*. A single insect, 16 to 19 mm is a queen of these species. The French-wasp worker - *Dolichovespula media* - is similar in size to the Common wasp queens, the *D. media* queen at 18-22 mm, is close in size to the worker hornet - *Vespa crabro*, whilst the *V. crabro* queen is the largest or our wasp family, 26 - 31 mm. The last two mentioned, have both spread northwards across Britain in recent times, but are quite different, the hornet being the most docile of our social wasps, bronze and black in colour, whilst the French wasp is easily our most vicious, and very varying in colour, from bright yellow striped to all black. Most black and yellow striped insects can up folk's blood pressure, there are beetles, moths, flies, caterpillars and a recent spider arrival - all harmless and expecting to be left alone - 75% of media photos depict a hover fly -*Eristalis tenax* - as a honeybee - one of my 'show' exhibits of two insects, entitled, 'Wasp or Hoverfly' even has beekeepers saying, 'Someone has switched your labels'! *Probably the single insect that causes most terror, is the Wood-wasp - Sirex gigas - I measured one at 41 mm to the end of her ovipositor.* The black and white ichneumon, predator of the Woodwasps, has also rated phone calls - a female can measure 100 mm from the tip of her antennae to the end of her ovipositor!!

Few folk will pick out the difference in colour - or shape - between honey bees and wasps, and if asked if they are rounded and furry, the best that many can do is pick out the queen bumbles. It does help if they can describe the Red Tailed Bumblebee, as they are mostly small nests and quiet - whilst the White and Buff Tailed can be larger nests and more 'active', even stroppy if disturbed. At least it is safe to tell the caller that the 'honey bees' flying from a 25 + mm hole in the ground are wasps - I only once found honeybees there; flying out of the finger-hole in the iron cover of a water stop-cock chamber - tell them the wasps are feeding flies and caterpillars to their offspring by the thousands, and that the nest dies in the Autumn, with only the young mated queens flying off to hibernate through the winter - no need to mention there can be a thousand young queens leaving a large nest! Of the five most likely species, I have found all of them, in all of the previously mentioned situations at one time or another. Nests are seldom in the same place the following year - but they do love dry compost heaps! If they are a danger, refer the caller to their local council or Yellow Pages.

Like wasps, bumblebees also over-winter as mated queens; explain that their small colonies do a great service pollinating fruit and seeds - often working in cooler conditions than honey bees tolerate! There are 22 species of bumblebee in Britain, some near extinction, and generally they are docile. The *Bombus lapidarius* nest ends quite early - sometimes before August - when the young queens will vacate the nest to mate and go into hibernation; like the Carders, Meadow, and Garden Bumblebees, they are mostly surface or above ground nesting. Even into June or later most nests are still quite small, with only dozens of bees at best, they are therefore quite easy to move; a four litre plastic ice cream container held down over the nest in the late evening, and a shovel pushed through the soil beneath, usually does the trick. One person can do it alone with say a brick holding the box down. String or parcel tape around shovel and box, and you are ready to transport it. Half a mile is adequate if it is placed beneath overhanging plants, so ensuring they realize the nest is in a new spot. Two bricks on edge and a slate or roof-tile on top, make a good shelter. I have never 'kitted up', for any of these bees. Often the nest is so small, a two litre carton is enough.

White and Buff Tailed bumble bees nests are mostly under ground. If disturbed by mowing, it takes bees - and wasps - some time to relocate because of the altered terrain, but once they have, they are no further trouble. It is helpful to move any material covering the entrance and mark it with a cane, then work can go on around them for the rest of the season - if much tall vegetation has been cut, and the hole not exposed, a few may never relocate, and will fly around aimlessly for a day or two. After marking nests of all species, I have worked all season close-by and never been stung, ever.

To move the larger, especially Buff Tail nests - I do kit up! I have used a twelve bottle wine carton, and a piece of plywood, cut to a sliding fit in the carton when on it's side. Pushed a shovel under the width of the nest, then slid the plywood beneath, picked it up and slid the lot into the carton If a nest has been much damaged; place the plywood in the box first, followed by some litter, arrange the cells as neatly as possible, keeping any honey cells upright, place the rest of the nest litter on top. The original nest area needs to be free of bees and debris so that the box can lay close to the ground - otherwise many bees will roost underneath; then leave them to settle in. Take a cloth cover when you move them the next evening; at the new site, slide the nest, complete with floor, into it's new sheltered position, cover with more dry grass or dead leaves.

I once saved a large nest discovered one metre underground during trenching using a hessian sack. I rolled down the neck to form a saucer and placed it on a square of wood cut to size, then arranged the nest as tidily as possible and put it all back in the hole. The next evening I quietly pulled up the neck tied it, and lifting it by the board base, took the sack a mile away and when the bees had quietened down, slit a hole in the side, and pushed a short piece of tube in to keep it open, then put some more dead leaves on top, and a shelter. This was an occasion when I was stung four or five times - by the same bee, for bumblebees live to sting again - but I found the pain much less than a honey bee sting.

Calls about solitary bees and wasps - over 400 species in Britain - are mostly at emergence, when both males and females congregate around the nest holes to mate. Days later only the females are in evidence as they dart in and out to gather cell construction materials or nectar and pollen, and if they are wasps, searching for prey They never attack; when handled, most of the bees only give a feeble sting, but the wasps are very variable - usually by size! They are an interesting lot; there are bald, brown coloured bees, some hardly a third of the size of a honey bee, hairy bees, from foxy red to golden yellow, drilling into paths, banks, plant-stems, dead-wood and walls in which to construct their cells. Others look for ready-made holes, even key holes and the tops of garden canes, some bees are so tiny they can construct three or four cells in a small snail-shell. The bumblebee-like, 'Hairy-footed flower bee' - the black female is usually being closely followed by the brown male - are rare except in areas of old chalk-clunch and clay-lump walls or lime mortar brick work, where they progressed from chalk-cliff faces. I have seen these blamed for disintegrating walls; however, of the many walls I have observed, only 1950s/60s, high percentage sandy mortar suffers - the holes collapse. Lime mortar walls that they have used for over two hundred years are still fine - holes are reused again and again.

Another group cut out discs from plants and flower petals from which to make cigar like rows of cells, placed either in holes dug in rotten stumps, or ready made holes, such as the end of a coiled garden hose, and dry sandy places, pots of cacti are ideal - gardeners mostly notice those that cut holes in rose leaves! There are even cuckoo bees who lay their eggs in other bees cells, and one or two, both wasps and bees, where the male stays around, even helping to guard! Solitary wasps use the same sites and methods, except I have not heard of a 'Leaf Cutter Wasp'. Many are in the familiar black and yellow livery - as are one or two of the bees - yet others are black and red - one black one is so small that it uses the 2 mm holes of the 'woodworm'. Most specialize in hunting one particular prey, aphids, hover flies, caterpillars, beetles, spiders and even other wasps. There are some 30 species of Ruby Tailed Wasps, each sneaking into a particular wasp or bee hole to lay an egg!

All wasps and bees will visit flowers to feed themselves, and in doing so pollinate - the social wasps do little flower visiting when there are grubs to feed, because of their method of taking the waste carbohydrates back from the larvae whilst feeding them their protein. You may answer your caller with, 'They are wasps', when you are told bees are spoiling their plums or grapes, although there may be bees there, it's the wasps who cut open the fruit in the first place! It is back at the nest, when feeding the grubs, that the two species are truly separated. Social wasps masticate and feed animal protein mouth to mouth to their pupae, solitary ones lay an egg on or with their paralyzed prey, whilst the social bees, after the eggs hatch, feed the pupae a puddle of protein manufactured from pollen and nectar or honey; and the solitaries lay an egg with either a ball of pollen or a puddle of nectar/pollen mix - like the solitary wasps they never see their babies! Don't expect to always get it right. One gardener's answers led me to believe there was a 'Media' wasp nest in the heart of a gooseberry bush; he was delighted when I said I could deal with it, as I was passing that way. Upon arrival, a very short sighted old fellow led me to a good honeybee swarm! If you are a fully paid up BBKA or affiliated Member you are insured for working with honeybees - Apis mellifera, and nothing else! Premiums keep on rising, simply because beekeepers are attending to bees in an increasingly urban environment - and folk are more likely to sue. It is obvious that Members should take into account the possible consequences of what they are about to do, it would not be considered very clever, if a dangerous situation arose, only because of a beekeepers intervention. Even if you have your own private insurance, once off of your own property, circumstances change. And most importantly, a BKA Member is not insured against third party litigation if someone suffers whilst the Member is messing about with bumblebees or wasps. And, by the way, smoke does not subdue bumblebees, wasps or honey bees without honey - stirs them up a bit though!

Any sensible beekeeper can do most of what I have described, and I have at least resisted explaining how I have moved wasp and hornet nests! It is more a case of 'Don't do as I do, do as I say'! And I must admit I have also on occasion talked beekeepers through a 'DIY' job. If like me, you cannot bare to see wildlife unnecessarily killed, and you enjoy a challenge, by all means have a go. I have described the methods I have used, dictated by always having to tear myself away from a busy life, so looking for the fastest, easiest, but still safest ways! If you like to be more meticulous and even enjoy the moment by putting them in glass observation boxes etc, do so by all means, just make sure that the inevitable by-standers know they could be in danger, and that you will not be held responsible if they stay and get stung. Also I should think we could be held responsible if someone gets stung by the inmates of a wild nest that we have relocated, also, if you are encouraged to demolish something to get at a nest, only do so, after, you have their instructions in writing! Complicated, ain't it!

It is some 70 years since I discovered that it is better to look a bit of a pansy arriving with all your beekeeping gear to gather a benign looking swarm, than look a fool when covered in a stinging mass! It is because of such incidents as this, that I know I am one of the lucky ones, my life is not threatened, even with multiple hundreds of stings! When some years ago, I heard that Addenbrookes Hospital wanted to know the whereabouts of wasp nests to milk the venom, I got in touch and became their main provider. It is strange that an insect coloured yellow and black as a warning to others, should be so hard to see during it's comings and goings. They needed to be in the ground and I found it was easiest to see the lines of flight when the sun is low, and depending on the background, either with the insects against the light, or with the sun behind you. It was not unusual to know I had found the nest, when I felt stings on my ankles!

When dealing with bees and wasps nests I would recommend you display a notice - even when working in your own apiary. I enclose a photo of mine - made up from 'The Bees of Swanland' by David Bone - yes I know they are stingless drones but it works a treat, and should at least save you the shock I once sustained! A hand clutched my shoulder as I opened a very stroppy hive and a voice in my ear said, 'Can I have a word, Mr Clar......?' Still, he was a Scout Master, and he very quickly remembered their motto, 'Be prepared'!

RECORD KEEPING
The Next Fifteen Years

Bill Bielby, C.B.I. for Yorkshire

They say that self criticism is good for the soul and having a good cry helps to make us feel better. If screaming would do any good to persuade beekeepers to keep detailed and accurate records of every single colony of bees then let BIBBA members howl from every hive roof for the next 15 years. We British Beekeepers are pathetic with our record keeping. Over a period of thirty five years of visiting and demonstrating at apiaries various NOT ONCE have I encountered any records whatsoever which would contribute to the selection and improvement of stock. In general, a national campaign is overdue: (a) to train for proper evaluation and assessment (b) detailed record keeping and (c) selection and breeding objectively from the previous five years of records. Opinions and word of mouth evaluations are OUT. '

Attitudes towards record keeping may be greatly influenced by schools. For example, some children (and teachers perhaps) would resent having to write up a daily account of the work done during that day but on analysis would this not show up the strengths and weaknesses of the whole system plus the eventual acceptance that recorded facts are firm foundations for improvements?

If the qualification for membership of British Isles Bee Breeders Association were that a beekeeper should have to produce evidence of having kept accurate records of his colonies' characteristics and performance for the last 5 years - BIBBA would have very few members indeed.

How can we overcome this national weakness? Only by sheer ruthless determination and persistence. By not being lazy. By being thorough and at the same time, not losing that national virtue - flexibility. But do write down. honestly as much as possible immediately a hive has been manipulated and let us steadily get rid of those awful bees to be replaced by thoroughbreds of known characteristics and performance.

from The Beekeepers Annual, 1983

We Need to Keep More Records!

Joss Langford

In recent years I have become increasingly aware that the record keeping suggested by mainstream beekeeping practices pushes new beekeepers towards a highly interventional style of management. I have kept my records in a free text manner and allowed the key information I need to emerge. In this article I set out some guidelines and design suggestions for an approach to record keeping that is both sympathetic to natural beekeeping and encourages mainstream beekeepers towards less intervention.

It would be unusual for a conversation about natural bee husbandry to start with an enthusiastic description of record keeping. The bees don't care whether or not we record their comings and goings. The bees have no use for our records; all the information they need is written into their environment. As well as understanding the locations of resources on which the bees depend, foragers know the impending weather well before it reaches the hive. The outside temperatures and changing day lengths allow the planning of brood expansion, while pheromones from the queen let the house bees know her productivity potential. The combs of the hive are always available to assess the status of stores, where every cell holds information about its use in the preceding years.

In contrast, we humans can maintain knowledge over many generations of different colonies. We have information over much greater distances, for example on the prevalence of diseases, and we have access to longer range weather forecasts. Most importantly, we can anticipate our own actions and ambitions in the stewardship of a colony.

The study of bees for science and research relies on the collection of data. This can be over many years, for example in the assessment of evolving varroa tolerance, or across different populations in different conditions at the same time, such as when assessing the potential impact of commercial insecticides.

In legal jurisdictions around the world beekeepers also have responsibilities for the reporting of disease and the movement of bees. All three (management, research and regulatory compliance) require us to observe our bees closely and keep records. Our challenge is to collect the information that we can use to assist the bees and meet our responsibilities without disadvantaging them overall.

When I started with bees I quickly realised that common approaches to record keeping did not represent the style of beekeeping that I wanted to follow and so decided to keep my notes in free text format. With a growing number of hives and sites, this approach eventually become cumbersome and I recently reviewed my records, comparing them with mainstream record templates that can easily be found online. My early notebooks contained all sorts of information but slowly reduced to just the points that, from experience, I had found useful later. Looking back there are clear patterns of useful information that do different jobs at different times of year. These patterns reflect the types of record commonly found in mainstream beekeeping – inspection, movement and hive records.

The most commonly available template is the inspection record and this fact alone demonstrates the reliance on intervention that infuses conventional beekeeping. In addition, the immediate assumption of most templates is that the hive must be opened for any information to be gathered on the state of the bees inside. These templates are also dominated by a rush to find the queen. Now, I enjoy spotting a queen, but it is more valuable to me and less intrusive for the bees if treated as a rare and privileged event. There is very little value in seeing the queen herself, the evidence of her recent actions and those of the colony provide a much better idea of her status.

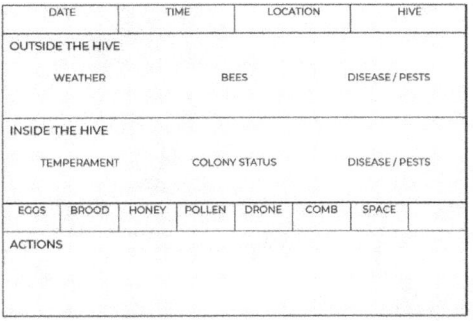

Table 1 shows a suggested inspection template for a natural beekeeper. The top line pinpoints the inspection in time and space. The next section prompts the beekeeper to consider the information around the outside of the hive and at its entrance (Figure 1). This template does not explicitly include what is, arguably, the most important matter when we consider interfering with a colony, the question 'why?'. The purpose of an inspection is something that I work-over in my mind every time I approach a hive, almost as if I were rehearsing for an interview. This 'outside the hive' section gives the beekeeper time to reflect on whether they already have enough information to complete the objectives of an inspection without disturbing the colony. If, in early season, the bees are flying well with over half bringing in pollen there is unlikely to be justification for further meddling.

The bottom sections of the template prompt key observations from a closer look at the hive. I keep top bar hives in three apiaries and with several smaller, tree-housed bait hives. For the bait hives, it is normally enough just to record whether they are occupied. In the apiary, top bar hives offer access to a significant amount of information with removal of just the follower boards. I routinely record the number of bars of new comb and remaining space to track colony build-up (Figure 2). Even just the total number of occupied bars provides useful year-to-year summary.

The focus of mainstream beekeeping inspections is often the attempt to control swarming. In common with most natural beekeepers, swarm prevention is not something to which I aspire, other than ensuring a colony has the space to expand. Nevertheless, knowledge about the reproductive status of a colony can be useful, and records permit a beekeeper to identify both the immediate trends and general nature of a colony. Understanding the likely behaviour of a colony assists with planning and, in turn, prevents unnecessary inspections.

The downloadable version of this template* also contains a notes section for infrequent events and those specific to a location, interest or management style – examples include the purpose of inspection, plants the bees are working, feeding, honey extraction or treatments.

Movement records are relatively simple other than where they are governed by regional legislation. Finally, the hive (or colony) record gives the opportunity to summarise essential information about a colony and its home. Updating hive records once a year from my inspection records helps me to plan my goals at the beginning of each new year. An example of a hive record is shown in Table 2.

DATE
HIVE / COLONY REF.
HIVE TYPE
ENTRANCE POSITION & DIRECTION
COLONY SIZE & STATUS
BEE SUBSPECIES & DESCRIPTION
ORIGIN OF QUEEN & COLONY
YEAR QUEEN MATED & SUPERSEDURES
FEEDING & HONEY REMOVAL
DISEASE & TREATMENT HISTORY
TEMPERAMENT
SWARMING HISTORY
COMB CONDITION
HIVE CONDITION
LOCATION (DESCRIPTION, ALTITUDE, HEIGHT, ENVIRONMENT, REF.)

An important aspect of record keeping is to decide the most appropriate format of your records – a practical and personal choice. Some beekeepers will prefer smartphone-based applications which have the advantage of instant digital storage but are difficult to configure for a particular style of management. I find that paper provides an affordance that I cannot forgo. I use a pocket-sized notebook which is large enough to use in the apiary (Figure 3) but small enough to be handy when checking more remote bait hives.

Those with many tens of colonies will probably look for a column version of an inspection template where multiple hives can be listed in a single table. The ideal option might be for a paper-based approach that can easily be digitally scanned with text recognition, but we have yet to find a workable system.

In summary, beekeepers need to maintain records for management, research and legal compliance. Keeping the right records should allow us to manage with less intervention. As in all human systems, the observations we request of a management approach will drive the management itself. We need to design record-keeping practices carefully to encourage the lowest required levels of intervention.

*Natural beekeeping templates for inspections records and hive records available for free download from Blackdown Bees (www.blackdownbees.co.uk).

- from Natural Bee Husbandry, No 8, August 2018

NATIVE BEES
History of the Native Honey Bee in Scotland
John Durkacz

It is postulated that the honey bee was pushed south by the advancing ice caps towards the Mediterranean region and extreme south western France during the last ice age. These surviving populations then slowly moved northwards following the receding ice as the climate warmed again and were able to cross the land bridge which existed around 8,000 years ago into what are now the British Isles. We can only guess how honey bees lived before man began to substantially alter the landscape by forest clearance. The Scottish climate and landscape has always been marginal for honey bees though the central and eastern areas were more suitable.

The central lowlands and eastern parts were the most fertile and there is evidence of beekeeping associated with the monastic houses and abbeys from the north east to the borders. The bee boles register (International Bee Research Association) shows the majority of bee boles situated in the east in those most fertile regions with lower rainfall; the largest number in Tayside, Fife and the Lothians. Many lesser-known bee boles were identified by beekeepers in the Fife and Dundee areas and it is likely that many will have disappeared by now. They are a testimony to the integral part that honey bees played in our rural economy in the past. We should not ignore the fact that bee boles of a much simpler design were also widespread in smallholdings in the north east and often went unrecognised.

Easter Lynne, Stratha'an.

Abandoned WBC and Cottage hive with roughly constructed wall with old bee boles for housing skeps. I found several of these in old farms and cottages in the north-east Highlands where I kept bees for many years. Many were sadly neglected and are likely to be lost by now.

Photo: J Durkacz

Beekeeping became an important part of rural activities during the 19th century when skeps were used, protected from our changeable weather in bee boles. The indications are that numbers of colonies fluctuated considerably depending on the seasons. Also many beekeepers were importing queens and small colonies from the continent in the belief these were more productive and in many cases easier to handle than the unimproved local bees. Even by the end of the 19th century there were comments by Scottish beekeepers that the true native black bee was becoming more difficult to find.

In the early years of the 20th century there were continuing importations of bees although it is likely that in remoter areas beekeepers continued to use the native bee, but there is also evidence of dispute about which bee was 'best'. Some were saying that anything was better than the common black variety and others strongly supported the native dark bee, voicing concerns about the wintering ability of imported bees. By 1912 the Isle of Wight epidemic had reached Speyside. The 'Scottish Beekeeper' reported a 'terrible year' with a 'wet and sunless summer' and whole apiaries wiped out.

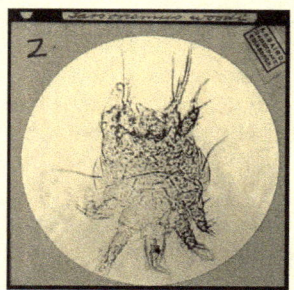

Original slide of the acarine mite or Tarsonemus woodi as it was then known. Dr John Rennie's work was funded by AHE Wood (Glassel House, Aberdeenshire) and his researches led to the discovery of this tracheal mite which was found in all the samples from colonies affected. Much later work suggested that transmission of viral disease was the cause of the devastation.

Photo: from old glass slide in SBA archives

After 1912 the IoW disease swept through the north on several occasions and around 80% of colonies were lost but there were survivors. Adverts in the Scottish Beekeeper show where replacement imported colonies came from but these were not all Italian as many dark bees were imported from France and Holland.

Scottish Beekeeper adverts.

During the interwar years beekeeping remained an important part of the rural economy and many experts in the Scottish Beekeepers' Association worked hard and travelled widely supporting beekeepers and helping in the transition from skep to modern beekeeping methods. The college advisory services played an important role here. Dr John Anderson in the North of Scotland College of Agriculture was a hugely important figure and under his influence the number of beekeepers rose dramatically. DM McDonald from Ballindalloch was just one notable expert who travelled widely in the North East supporting beekeepers through these difficult times. But later, in the more distant areas, the number of beekeepers fluctuated and declined as depopulation of the rural areas began. I managed to interview two elderly farmers in the Stratha'an area many years ago. They were Mr McArthur from 'The Knock' and Paddy MacDonald from 'Inverchor' and both had small commercial beekeeping interests in their younger days.

Paddy MacDonald had around 100 colonies using Italian bees in the low-lying areas and black bees near his hill farm which was in a remote area. He had his own waxed cartons for run honey and specialized in comb honey.

Photo: J Durkacz from original documents

Mr McArthur's apiary in Stratha'an after he gave up active beekeeping. One of these hives had a strong colony of dark native bees which I was able to transfer.

Photo: J Durkacz

Moving to the Borders we find major beekeepers running commercial enterprises successfully who did so using local native bees which they were able to select and improve. Willie Smith from Innerleithen began his beekeeping after the 1st World War, eventually running about 150 colonies as a single-handed, full-time commercial venture. In the later 1920s he developed a single-walled hive with short frame lugs and top bee space later to become the 'Smith Hive' which was widely adopted in Scotland. He was probably the first full-time commercial beekeeper in Scotland and much revered and his advice sought after. He used locally-sourced native bees as did later beekeepers in the area, Willie Robson from Chainbridge and George Hood from Ormiston (East Lothians).

Willie Smith, probably Scotland's first full-time commercial beekeeper. After he retired from beekeeping many of his hives went to George Hood and others to local beekeepers. Subsequent surveys have confirmed a higher concentration of native dark bees in the borders area.
Archive photo: permission of Morna Stoakley and Amanda Clydesdale

In the 1970s and 80s in the north at Craibstone, Bernhard Möbus became a popular beekeeping advisor and champion of our native honey bees. Bernhard wrote widely on the effects of Varroa (which at that time had not reached our shores), bee breeding and the use of small mating nucs. He also researched wintering abilities of honey bees and the effects of winter brood rearing and presented his papers at Apimondia. He recognised the need to have a bee that was adapted to Scottish conditions, was docile and could winter well. He discovered colonies belonging to an elderly beekeeper in the village of Maud (Aberdeenshire) which fitted the required characteristics and raised numerous queens which were spread around Scotland. Descendants from these colonies are still to be found.

John and Morna Stoakley were keen beekeepers who moved to Scotland in 1968. They soon discovered that local bees were the best for their apiaries near Peebles and were inspired by Bernhard Möbus' teaching. They had strong connections with the SBA and BIBBA and started bee improvement classes teaching BIBBA developed selection procedures for the native honey bee. These early classes were held in a rural primary school in Stobo and attracted excellent teachers through BIBBA for wing morphometry classes. They were also involved with the organisation of the joint SBA/BIBBA 1992 conference 'Northern Bees in the 90s' held in St Andrews. John Stoakley was an entomologist by profession working for the Forestry Commission and well versed in procedures for surveying insect populations. By 1994 they had completed a wing morphometry survey from apiaries in Scotland which showed that a third of colonies were native honey bees, a third were 'near natives' and the remainder hybridised. This confirmed that there were significant surviving populations of native dark bees despite continuing importations.

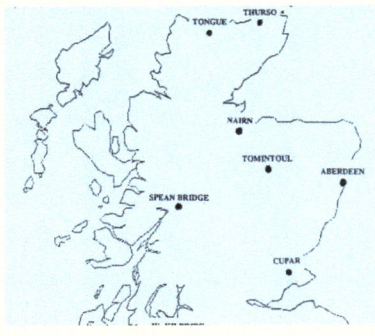

Colony samples were taken from apiaries including Peebles and Central belt not shown on this map. The apiaries had around 10 colonies each and the sampling was done randomly from these apiaries. Measurements were made manually for discoidal shift and cubital index and plotted to show distribution.

$1/3$ of sampled colonies were native types
$1/3$ were near natives
$1/3$ were clearly hybridised

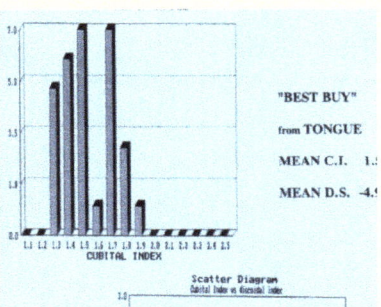

Example of plotted result from the original survey (1993-4).
This colony from Tongue had a low cubital index and marked negative discoidal shift.

The untimely death of John Stoakley and the closure of Craibstone with the departure of Bernhard Möbus were serious blows for the conservation of Scottish native honey bees but the 'Maud' strain had already been widely dispersed. Bernhard wrote about his work in the college journal where he described the rearing of his preferred strain, and there are confirmatory articles in the 'Scottish Beekeeper' explaining how associations visited Craibstone and many young queens were collected by members.

A more recent survey involving samples from another survey (Ewan Campbell, Aberdeen University with Jim McCulloch) where DrawWing was used for morphometry checks showed very similar results to those from the Stoakley survey.

Despite the even heavier importations of foreign bees over the past few years, there is good evidence that the native bee survives in Scotland. Europe-wide studies confirm that locally adapted bees survived longer. The native bee is able to respond more readily to local forage and climate changes, holding back on brood rearing, storing large amounts of pollen and has better wintering abilities. This enabled survival during periods of neglect in beekeeping when economic depression led to depopulation and a reduction in the number of beekeepers in many areas of Scotland.

John Durkacz

Visual assessments to identify Native dark honey bees, *Apis mellifera mellifera*

Gavin Ramsay, Chairman

John Durkacz, Member

Scottish Native Honey Bee Society

With experience, it is possible to identify good candidates in searches for the native dark honey bee of the Western fringes of Europe.

Body colour should be uniformly dark brown. Small lighter tan spots at the sides of the 2nd tergites may not indicate hybridisation.

Body shape is useful. The Carniolan and Italian races have a slimmer appearance with a more pointed abdomen.

Thoracic hair colour is a good indicator of bee type. The area on the upper surface of the thorax has black and brown hairs in the dark native honey bee. Carniolans, Italians and Buckfast are paler, usually with no black hairs. The 'halo' of pale hairs around the thorax is very noticeable in the eastern types of honey bee and less prominent in dark native honey bees. Allowance should be made for the effects of backlighting.

Tomenta are the bands of short, fine hairs running across the abdominal segments and most easily seen on the **visible** 3rd, 4th and 5th tergites shown in the diagram below. Although they can be accurately measured, this is unnecessary for our purposes. A visual assessment of the widest part of the hair band across the **visible** 4th tergite comparing with the rest of the tergite seen towards the tail end is all that is required. **There are three questions which serve to discriminate native honey bees from other types:**

1) **Is the hair band equal to, less than or more than half the width of the distal part of the tergite?** The native honey bee has a narrower tomentum. If the tomental band width is 40-50% it would be classified as medium and more than 50% as broad. Some may find it easier to use a low-powered hand lens to assess this. Broad tomenta are not seen in dark native honey bees.

2) **Are the hairs a little sparse?** Good examples of Carniolans, Buckfast and Italians show denser hairs and a bolder, paler appearance of the tomentum. Bear in mind that young bees are covered in pale hairs which disappear as they mature.

3) **Comparing the 5th visible tergite to the 4th, is the tomental band markedly narrower?** The eastern types of honey bee have tomental bands on the tergites that are more equal in width across the different tergites and are more uniformly broad.

Assessing whether a sample of bees fits the dark native honey bee for tomentum bands involves a consideration of these three features. In practice, all of this can be assessed at a glance.

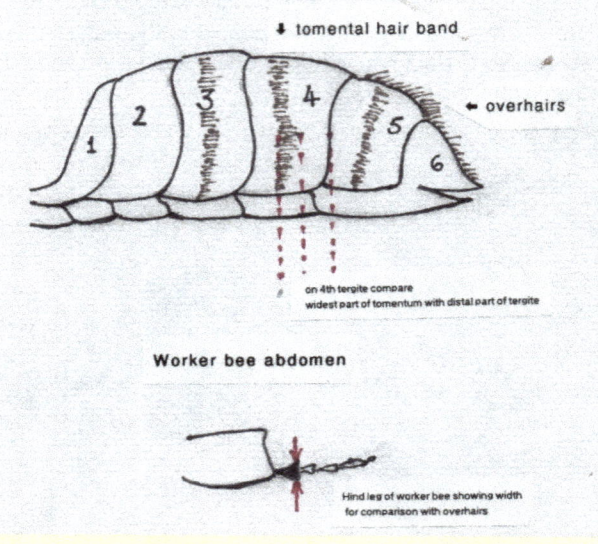

Overhair length can be assessed by comparing the **width of the first foot limb of the back leg of the bee which is about 0.4mm.** (Blue arrow below). In the photo a quick look shows hair length on the 5th visible tergite (red arrow) against a lighter background is at least the width of the first foot (0.4mm). The dark European honey bees can have hairs clearly as long whereas Carniolans, Italians and Buckfast will have short, neater hairs.

As with all visible features, expect a pure stock to be fairly uniform for all these traits. Young bees with extra pale hair and worn older bees should be avoided in a visual assessment. Variation not accounted for by the age of the bee suggests the colony is hybrid to some extent.

WORKSHOP
Build the Original Langstroth Hive

Peter Sieling
7201 Craig Rd, Bath, NY 14810 USA

Langstroth's 1852 patent moveable comb hive was the first practical hive that allowed the beekeeper to examine any comb in a hive quickly and with minimal disruption to the bees. His original design, patented in 1852, looks complicated, but it's actually simple to build, requiring only a few hand tools and a table saw. Langstroth assures us the saw can be powered either by steam, water, or horse power. If none of these power sources are available, electricity will work fine. Besides a table saw, you'll need a drill press with 1 1/4" and 1 3/8" bits, and a router with a straight cutter.

Langstroth's plans are not so easy to follow. He tended to leave out details or put them somewhere else in his book. The illustrations and descriptions don't quite match. He offers five hive models, from a basic box to a complicated double walled hive with a glass window. He recommends his "hive #2 without observing-glass" for "those largely engaged in bee culture". This is the model described here.

Langstroth's hives were meant to be stacked. The lower box contains the brood; the upper box was for surplus. Unlike the modern hive, you stack an entire hive, including the bottom board, on top of the lower hive. The bottom board has holes drilled in it allowing access to the upper hive. You can make multiple hives and stack them, or just put a modern super on top.

Lumber in the nineteenth century was planed to 7/8". Almost all components are based on this thickness. You won't find this at building supply centers, but any lumber company selling rough lumber can custom surface lumber to 7/8". Langstroth recommends cedar, basswood, poplar, or pine. Other species will work as well.

Constructing the Box

1. Cut two sides 10 7/8" x 23 7/8". Cut the front to 8 7/8" x 14 1/8", and the back to 8 7/8" x 15".
2. The sides are mirror images of each other. On the sides cut a 2 1/8" x 4" notch on the upper front (see diagram). Mill a 7/16" groove across the bottom of the sides, 7/16" above the bottom edge. The floor fits into this groove.
3. The back will be "halved into" the sides. Cut a 7/16" x 7/16" rabbet on the ends of the back. Cut a mating rabbet in the back of the sides. This rabbet starts 5/8" from the top of the sides and stops at the lower rabbet. Nineteenth century carpenters would have bored a row of holes 7/16" deep, and then chiseled out the waste wood by hand. Using a router with a straight bit is easier. Clamp a straightedge to the side to guide the router. Square the upper end with a chisel. A simpler but weaker construction would be to skip the rabbets, cut the back to 14 1/8" and nail in place.

Floor

4. Both floor and roof are tongue and grooved, so mill them both at the same time. The grain on the floor runs from side to side, the roof runs end to end. After tongue and grooving, crosscut to 15" enough material to make a floor 23 7/8" wide. Mill a 7/16" rabbet on the end grain. Make sure it slides easily into the groove on the sides without play. The bottom of the sides should be flush with the bottom of the floor.
5. When hives are stacked, the bees pass between stories through six holes in the bottom board. Start with the 1 3/8" bit, drill 1/16" deep, then drill through with a 1 1/4" bit. For the lower box, close the holes with 1 3/8" disks cut from tin. When using a second hive for surplus honey, remove the disks.

Portico Roof and Top Rails

6. Langstroth's hive's distinctive look comes from the portico, the porch that extends 4" past the hive front. Cut a board to 4 1/2" x 17 5/8". Bevel the front from 1/2" to full thickness 2 1/2" back, then round the front drip edge.
7. The frames rest on top of the two ends. The front and back rail form the back of the frame rests. Both rails are 15 7/8" long. The back is 2 1/8" wide. The front, which sits on the portico roof, is 1 1/4" wide. The front rail has an upper entrance, 3" long by 1/4" wide and slanting downward from inside to outside.
8. Make two ledges to run from the portico roof to the back rail, 7/8" square by 20 3/4". One back ledge, 7/8" square by 17 5/8" completes the hive body.

Triangular Entrance Regulators

9. Make a block 4" x 5 3/4". Cut in half diagonally. On the bottom, cut grooves 1/8" deep x 1/2" wide. The grooves are supposed to attract wax moth larvae for you to remove periodically. The entrance regulators remain loose and can be added or removed as needed, depending on the strength of the colony.

Cover
10. Make a panel from tongue and groove stock 25 3/4" x 19" with the grain running lengthwise. Langstroth recommends rain-grooves at the tongue and groove seams—bevels forming a gutter to draw water away from the seams. A better solution would be to cover the top with sheet metal. Two cleats, 2" x 19", fasten to the underside, flush with the ends.

Frames
11. Make ten top bars 5/16" x 1" x 19 1/8". If you plan to fasten comb to the top bar, you are done.
12. If starting a colony without combs, you'll need triangular comb guides. Crosscut the wood to 16 3/8". Set the table saw arbor to 30 degrees and the fence 7/8" from the bottom of the saw blade. Use a push stick to prevent kickback. The first cut will be a waste piece. Rotate the stock end over end and saw out the first triangular piece. Continue until you have ten per box.
13. Make twenty side bars 1/2" x 7/8" x 8 5/8".
14. Make ten bottom bars, 1/4" x 7/8" x 17 3/8".

Moveable Hive Stool
15. Make a front and rear leg, both 20" long, the front 5" wide and the rear 7".
16. Make two pieces 1 3/4" x 32". Nail them across the top of the two legs as illustrated, projecting 4" beyond the rear leg and 9" beyond the front leg.
17. Two more pieces, 1 1/2" wide are trimmed to fit and attached to the sides to brace the legs.
18. Hem a piece of cotton duck canvas, 8" x 20". Tack to the front as an alighting board for the bees. The hive fits loosely on the stand. Langstroth's illustration shows small wooden wedges holding the hive in place.

Assembly
Use exterior nails, zinc coated or stainless steel for nailing the hive together. For maximum strength, use spiral threaded nails, 2"–2 1/2" long when nailing into end grain. Use 1 1/2" nails when nailing 7/8" to 7/8" pieces. Drilling pilot holes just smaller than the nail's diameter prevents splitting and increases a nail's holding power.
19. Nail the floor to the sides. The back end should fit into its grooves and flush at the back. The top of both ends should be 5/8" below the top of the sides and 3/8" above the floor. The front end is flush and square with the notch for the portico roof. Nail the ends in place.
20. Center and nail the portico roof. Nail the front and back rails. Add the two sides and one back ledge.
21. There is a back entrance at the bottom. Make a shim to fit (not pictured) and slide it into place. Most of the time that remains closed.
22. When assembling the frames, use 1 1/2" pneumatic staples or standard frame nails to fasten the top and bottom bars to the side bars. If making several hives, make a jig to align the parts for nailing.

23. Assemble the stool. Tack the canvas to the frame after painting.

Once the hive and frames are assembled, prime and paint the hive with exterior paint. The hive is ready to stock with bees. For directions on stocking and using the original patent hive, read Langstroth's Practical Treatise on the Hive and the Honey-bee. It is still in print or available online.

Materials
Hive Body

Part	Description	Finished size (inches) 7/8" thick except where otherwise noted	No. Required
A	Sides	10 7/8 x 23 7/8	2
B	Bottom board	15 x 23 7/8	1
C	Hive front	8 7/8 x 14 1/8	1
D	Hive back	8 7/8 x 15	1
E	Portico roof	4 1/2 x 17 5/8	1
F	Top front rail	1 1/4 x 15 7/8	1
G	Top back rail	2 1/8 x 15 7/8	1
H	Side ledges	7/8 x 20 3/4	2
I	End ledge	7/8 x 17 5/8	1
J	Entrance regulator	4 x 5 3/4	2
K	Cover	19 x 25 3/4	1
L	Cover cleats	1 3/4 x 19	2

Frames

Part	Description	Finished size (inches)	No. Required
M	Top bars	5/16 x 1 x 19 1/8	10
N	Comb guide (triangular cross section)	7/8 x 7/8 x 7/8 x 16 3/8	10
O	Side bars	1/2 x 7/8 x 8 5/8	20
P	Bottom bars	1/4 x 7/8 x 17 3/8	10

Moveable Stool for Hives

Part	Description	Finished size (inches) 7/8" thick except where otherwise noted	No. Required
Q	Rear leg	7 x 20	1
R	Front leg	5 x 20	1
S	Upper rails	1 3/4 x 32	2
T	Leg brace	1 1/2 x 30	2
U	Duck canvas	8 x 20	1

Langstroth Hive Exploded View

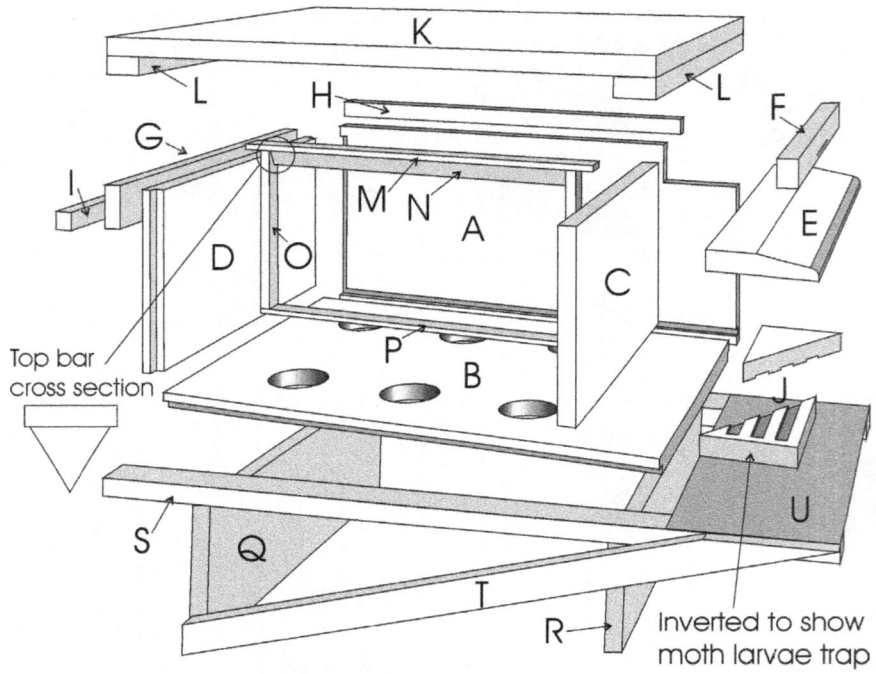

Note:
It took me approximately 40 hours to decipher Langstroth's plans and the result was surprisingly simple. I would make only some minor changes. I just learned that the frame length is 1/4" shorter than modern Langstroth frames because A.I. Root misread the dimensions. By the time he figured it out, they'd sold thousands of Simplicity hives so it was too late to change. Langstroth thought Root's frames should be called Langstroth Simplicity frames to distinguish them from the original.

1. Rabbet on Sides

2. Sides and Bottom Front View

3. Sides and Bottom Rear View

4. Fastening Ends to the Sides

5. Completed Hive Front View

6. Completed Hive Side View

Jörg Ruther

APITHERAPY
Long Forgotten Remedies for Natural Healing
Jörg Ruther, Germany

Apitherapy (from Lat. apis – bee) is the general name for methods of treatment of various human diseases with the use of (live) bees and bee products. The main products used in apitherapy for healing are honey, pollen, propolis, royal jelly, drone homogenate (drone milk), dead bees, bee pollen, beeswax and bee venom. People make ointments, tinctures and tablets from bee products as well as using bee stings in therapy. Besides the well-known bee products, modern medicine is beginning to discover a number of other rare and nearly forgotten products already in use by apitherapists in eastern and oriental countries. For centuries this knowledge has remained the property of a very small number of people. Beekeepers and healers carefully guarded their recipes and knowledge, passing the secrets down from parents to children.

Honey
Honey, besides being a food of bees is universally used as an agent for protecting human health. Its mechanisms are simple but very effective as it is based on the different combinations and interactions of carbohydrates and enzymes.

A widely disputed explanation of the term "honeymoon" claims that it comes from a tradition found in a number of cultures - German, Scandinavian or Babylonian, where mead was drunk in great quantities at weddings and then after the ceremony nuptial couples were given a month's supply of mead. It was believed that by faithfully drinking mead for that first month, the woman would "bear fruit" and a child would be born within the year.

There is a wide range of medical uses of honey, particularly as it has antibacterial properties and can be used as a remedy for burns, coughs etc.

Honey - the most widely used hive product which is very important for human health. Best to eat it in its purest state, straight from the comb.

Oxymel: A Medicinal Drink

A common medicinal preparation that dates back to antiquity is Oxymel, which is basically a mixture of honey and vinegar. It has many uses, and is a part of many traditional medicines. I remember my grandmother drank it through the hot seasons.

The simplest recipe to prepare Oxymel is to mix together 4 parts of honey with 1 part vinegar. Apple cider vinegar is the kind that is most commonly used.

Another recipe suggests that you boil a mixture of one part of vinegar, one part of water and two parts of honey. This mixture has to be simmered down slowly until only about a third of its original volume remains. While boiling it down, skim off any scum or froth that rises to the surface. I do not like this recipe as it overheats the honey and you might lose its good properties.

The author of a bestselling book in 1958 called "Folk Medicine" touted honey and apple cider vinegar as a panacea or cure-all. In this book ancient Oxymel preparations are mentioned to treat arthritis, gout, high cholesterol levels, as a metabolic stimulant to promote weight loss, and for longevity and life extension.

Some recipes add garlic to the Oxymel as an additional natural antibiotic.

Propolis

Inside the natural hive there is a perfect bacterial balance and some kind of sterility. An important role in maintaining the cleanliness in the hive belongs to one of the most valuable and unique products of bees: propolis or bee glue. Due to its antibacterial properties it protects the bee hive from germs, bacteria, fungi, and even greater dangers such as mice or lizards (which are mummified, preventing them from rotting within the hive). Fragrant and pleasant by taste, propolis is still a mystery even to the most experienced researchers. The chemical composition of this substance is very complex and varies from hive to hive, from district to district, and from season to season. Propolis is sticky at, and above, room temperature (20 °C). At lower temperatures, it becomes hard and very brittle. It has nearly 300 different compounds, but only around 100 of them have been identified so far. Northern European propolis has approximately 50 constituents, primarily resins and vegetable balsams (50%), waxes (30%), essential oils (10%), and pollen (5%). Propolis also contains persistent lipophilic acaricides, a natural pesticide that deters mite infestations. It is impossible to list all the benefits of propolis. Its most important quality is the ability to purify and regenerate cells, so that they are better able to rid the body of free radicals and hyperoxides. Scientists call propolis a "antioxidant bomb" and studies have shown that regular use of this product does indeed "rejuvenate"!

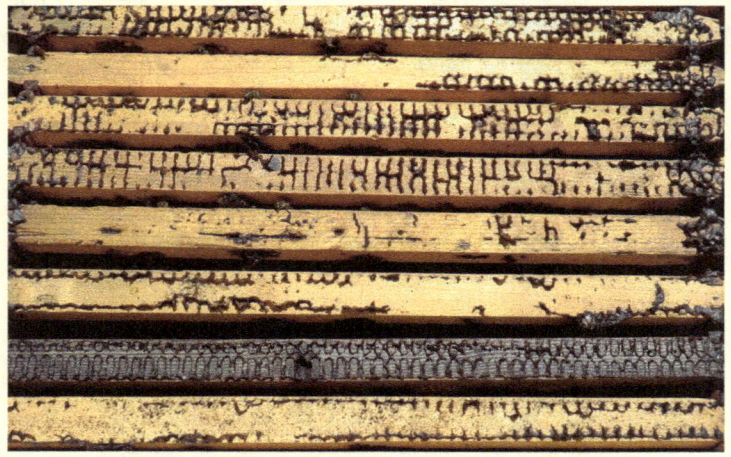

Propolis on the top bars of frames. It is a precious material and shouldn't be wasted.

Royal Jelly

Royal jelly is a honey bee secretion that is used in the nutrition of larvae, as well as adult queens. It is secreted from the glands in the hypopharynx of nurse bees, and fed to all larvae in the colony, regardless of sex or caste. Royal jelly is 67% water, 12.5% protein, 11% simple sugars (monosaccharides), 5% fatty acids and 2–3% 10-hydroxy-2-decenoic acid (10-HDA). It also contains trace minerals, antibacterial and antibiotic components, pantothenic acid (vitamin B5), pyridoxine (vitamin B6) and trace amounts of vitamin C,[2] but none of the fat-soluble vitamins: A, D, E or K. Major royal jelly proteins (MRJPs) are a family of proteins secreted by honey bees. The family consists of nine proteins, of which MRJP1 (also called royalactin), MRJP2, MRJP3, MRJP4, and MRJP5 are present in the royal jelly secreted by worker bees.

Royal jelly stimulates growth and development, increases vitality, stimulates metabolism, and normalises the endocrine system. The result has a powerful positive effect on all the body systems. Royal jelly has a great impact on our immune system. However, care is need with this and any other hive product as there have been documented cases of allergic reactions, namely hives, asthma, and anaphylaxis, due to consumption of royal jelly.

Dead bees - Podmor (Подмор)

Beekeepers collect dead bees, especially in spring when the amount and quality of the bees are at their highest. It is important that the bees are clean - i.e. without any signs of mildew or decomposition, and that they come from a beekeeper who does the not expose his bees to chemical treatments. Properly dried bees have a specific odour. Some say that it is sweet, slightly similar to the flavour of dried fish; others compare it with the smell of roasted pancakes or fried sunflower seeds. When you have once smelt it, you will never forget it; it is a very distinctive smell! The most valuable bees those are the ones collected in summer or autumn, because at this time the body of the bee has accumulated a huge amount of useful substances. In the summer months, bees have the opportunity to eat fresh pollen and nectar; they are more energetic and healthier. It is best therefore to use podmor from these bees for internal use and those from the end of winter for external application. The bees should be dried in an oven at a temperature not exceeding 50 degrees Celsius, after which they should be stored in a tightly closed box, a glass jar or a linen pouch. Another storage option is in the fridge freezer. Like other bee products, dead bees can be bought in Russia from beekeepers as they are is often sold at the honey fairs.

The effectiveness of podmor is undeniable and has been confirmed throughout history by practitioners of herbal medicine. Galen, the famous doctor of ancient Greece, used crushed honeybees against carbuncles, gum disease, toothache, to facilitate teething in infants, and to restore hair growth. The Roman writer Pliny Jr. pointed out that ash of burnt bees, mixed with oil, is a good remedy for many ailments including as an ingredient in an ointment for eye diseases, as well as for the growth and strengthening of hair. Bees, cooked in honey, were given to alleviate dysentery, and a broth made of bees in the form of tea was used as a diuretic. In 16th Century English medical books, dried and powdered bees were recommended for internal use, mixed into milk and wine, against dropsy, gout, rheumatism and as a solvent of urinary stones. Bees in honey would also help against spasms and stomach ache, bloody diarrhoea, and for the treatment of ulcers. The medicinal properties of podmor can be amplified and enriched by skilfully combining it with other biologically active bee products, as well as with herbs, fruit and vegetable additives, vegetable oils and other natural products.

The tincture contains all the substances you find in living bees such as apitoxin, chitin (chitosan), etc. Chitosan is a very useful substance, extracted from the chitin which covers the exoskeleton of bees. Chitosan has many properties that make it attractive for a wide range of use in food, cosmetics and skin care. Its use is very effective because of its film-forming and anti-inflammatory action. Another invaluable feature is its ability to sorption (the taking up and holding of one substance by another). The bees chitinous cover has a protective function, protecting the internal organs from the penetration of all kinds of pathogens. Crustaceans and even mushrooms have such chitin shells. Chitosan has a long history for use as a fining agent in winemaking.

Due to my classical education in alchemy, my approach to producing medicines is of a spagyrical* nature. Spagyric medicine is an herbal medicine produced by alchemical procedures. These procedures involve fermentation, distillation, and extraction of mineral components from the ash of the plant. This means that you make, for example, an alcoholic extraction, with the volume of the alcohol not exceeding 70 %, to make sure to dissolve the water-soluble components as well. The residues of this process, when collected after the alcoholic tincture is filtered, are burned to white ash. This ash will also be disolved in water, filtered and the water containing the water-soluble mineral nutrients and micronutrients are mixed together with the alcoholic tincture.

There are many solvents that can be used for apitherapy or in this case spagyrical* medicine. The bees are dissolved either in alcohol, in water (like a tea), in oil, in honey or in vinegar. The right vinegar would here be honey vinegar made from mead. You also can mix the podmor with water and honey, add yeast and make a kind of mead. You then either consume the mead or distil it and process the residues, as mentioned above, i.e. after burned to ash.

Perhaps the most common form of application in Russia is as a tincture (alcohol extract). For this purpose a glass of vodka takes one or two tablespoons of dried ground bees. To grind the bees it is possible to use a mortar and pestle, the raw material then being passed through a sieve. I also use a cheap electric spice and coffee grinder with stainless steel blades. The ground bees and alcohol are put in a jar which is left in a dark cupboard - I leave a jar or a bottle for three weeks. During the first week the mixture is shaken every day, afterwards just every two days. After the specified period the tincture is filtered through two layers of a dense bandage or gauze. The tincture can be taken three times a day; 20 drops in 100 ml of water. The taste is pleasant, but you can add a little spoonful of honey or a few drops of propolis infusion to the glass to enhance the positive effect. This dosage is suitable for middle-aged people. However, for elderly people it is recommended that they should drink daily as many drops as they are old in years, dividing the number of drops for two or three doses. Anyone weighing more than 60 kilograms should have the number of drops increased by 7 for every additional 10 kg.

After following this course of treatment even for two months will have a positive effect on a person's health. In general it is recommended to repeat the course of the tincture several times during the year. Its benefits are known to improve those people who have a weak immune system, heart diseases, problems with brain vessels, epilepsy, eczema, psoriasis, myoma, ovarian cyst, impotence and frigidity. The elixir improves the mood and gradually normalizes weight.

In diseases of the thyroid gland, especially in the detection of cystic formations, the podmor tincture is mixed with 10% tincture of propolis, with treatment recommended for three months. The tincture can also be rubbed into joints before bedtime to relieve arthritis.

There are many other well-known ways to use podmor in Russia but I would rather leave it to a doctor to explain them. But you should keep in mind that you also should change your diet so that it is more balanced. Stay away from alcohol and meat when doing a podmor cure.

Recipes:
Purpose: Joint and muscular pain
3-4 tablespoons of ground podmor. The ground bees are mixed together with oil. It is best to use olive oil or linseed oil. Try mixing it up to a creamy constituency, it must not be liquid. After the creamy condition is reached, place it in a dark closed bottle and store it in the refrigerator at a temperature of 5-7 degrees. When using, the crème is heated slightly and applied to the respective place.
Purpose: High blood pressure, for good blood circulation, strengthening the immune system, heart and kidney problems.
Pour 1 glass of podmor (important on this occasion that the bees are not crushed) and 0.5 litre of vodka into an enamel pot and cover with a lid. The pot is slightly heated for 10 minutes. Don't let it boil! Once cooled, pour the whole contents into a dark glass and allow to stand for 10 to 12 days, though give the mixture a vigorous shake every day. When the steeping time is over you can consume 1 teaspoonful per day. You now can filter the mixture but you don't have to.

Of course you should not use any of these medications if you suffer from a bee allergy! Always consult a doctor before you start any self-medication!

*Spagyric most commonly refers to a plant tincture to which has also been added the ash of the calcined plant. The original rationale behind these special herbal tinctures seems to have been that an extract using alcohol could not be expected to contain all the medicinal properties from a living plant, and so the ash or mineral component (as a result of the calcination process) of the calcined plant was prepared separately and then added back to 'augment' (increase) the alcoholic tincture. The roots of the word therefore refer first to the extraction or separation process and then to the recombining process. These herbal tinctures are alleged to have superior medicinal properties to simple alcohol tinctures, perhaps due the formation of soap-like compounds from the essential oils and the basic salts contained within the ash. In theory these spagyrics can also optionally include material from fermentation of the plant material and also any aromatic component such as might be obtained through distillation. The final spagyric should be a re-blending of all such extracts into one 'essence.'

The Process of Making Podmor:

1. Setup.

2. Dead bees in a coffee grinder.

3. The ground-up bees in the bottle.

4. The ground-up bees in self-distilled apple Brandy 57%, which will be ready in a couple of weeks.

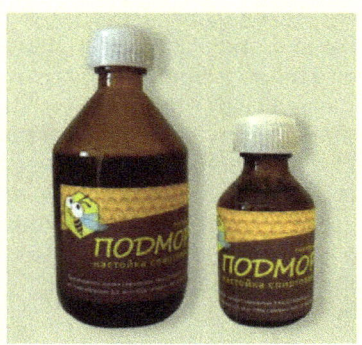

5. Podmor Products

Drone homogenate (drone milk)
Drone homogenate (drone milk) is produced by sacrificing drone larvae. This natural product was used in ancient times. In a tomb dated from the Han dynasty in the province Huang (China) recipes were found written on bamboo with a description of the use of drones' milk. It is a thick creamy liquid, with grey-yellowish colour and peculiar taste. Drone homogenate is a precious bioactive product. It includes: proteins, vitamins, ferments, microelements (calcium, magnesium, iron and etc.) and is an excellent source of phytosterols. Phytosterols are antagonists of cholesterol compositions and have immune modulating and anti-oxidative activity. According to research, drone homogenate bioactive property is higher than the one of royal jelly. Proteins of homogenate are presented as 37 amino acids and easily absorbed transport oligopeptides. Drone homogenate also includes 30 unsaturated fatty acids, water soluble (B group) and fat soluble (A, D, E) vitamins. Drone homogenate loses its useful properties easily because of environmental factors, so it's necessary to take this product, conserve and stabilise immediately.

Drone milk is also known as "Apilarnil". This product got its name, "Apilarnil", from its inventor - the Romanian beekeeper Nicolae Iliesiu. One day, about 30 years ago, his neighbour reported about his - in contrast to earlier years - remarkably fast-developed duck breed. The beekeeper Iliesiu had given the cut out drone-combs to the duck chicks to pick out. After this observation, he has repeated the experiment with the result that the chicks that got drone larvae grew faster than the peer group. From this he concluded that the drone-lavae must be the reason for the good growth of the young ducks and thus discovered the great potential that lies in their use.

Wax
Wax is able to maintain its plasticity and anti-bacterial properties for centuries. Pieces of wax found in the Egyptian pyramids were still soft and had a characteristic odour. Its ability to preserve products is utilised in wine-production, as the inside

of wine barrels were covered with wax. The ancient Egyptians used wax for mummification, in preference to more affordable Dead Sea salt, also known for its antibacterial properties. With its antibacterial properties, wax seals the cells of pollen so it does not become mouldy despite the high humidity in the nest, as it completely isolates the product from the influence of the environment. Caps cut from sealed honeycombs, called zabrus by beekeepers, differ from conventional wax used in the hive, as bees use a special substance, which includes mixed secrets of wax and salivary glands of bees, propolis and pollen and it is much better absorbed than regular wax. Sealed honey never spoils. Wax is mainly used in cosmetics and medicine to make salve and lotions.

Zabrus -ЗАБРУС
Zabrus is the cut strip of the upper lids of sealed honeycombs. It differs from the rest of the wax in the hive. Studies have shown that zabrus is highly effective in the treatment of bacterial and viral diseases of nasopharynx and upper respiratory tract. Chewing of zabrus is useful in many ways as it causes severe salivation, which increases the secretion and gastric motility.

Doctors recommend chewing one tablespoon of zabrus for 5-10 minutes, 4 times a day. It is easily digested with a dual benefit for mucosa and teeth.

Zabrus, the fresh wax cappings of honeycombs, should be collected and bottled for a wide variety of health problems.

Bee pollen - ПЫЛЬЦА
Bee pollen is obtained from pollen collected by bees during pollination. It also contains small amounts of bee saliva and nectar. Its nutrient composition can vary from region to region depending on the plant species visited. Bee pollen has been used by various cultures for centuries and is considered a superfood due to its high nutrient content. Nevertheless, pollen is not suitable as the sole source of nutrients for humans, as their vitamin C, D and B12 content is insufficient.

However, pollen can contain a lot of iron. In 1991 a group of scientists put forward their findings of the benefits of pollen: "Properties of DNA-reducing nucleases". The gist of it is that special enzymes in pollen penetrate the cell nucleus and during the synthesis they restore the damaged DNA chain. As a result, the life of the cells is extended several times. Bee pollen can be used as anti-allergics, a remedy for degustation problems and as a support for strength and endurance.

Pollen in the comb and collected for drying.

Perga - ПЕРГА

What beekeepers call perga, or ambrosia, or "bee bread" is a pollen ball packed by worker honeybees into pellets. Bee bread is field-gathered flower pollen stored in brood cells with honey bee saliva, sealed with a drop of honey. With the leaf-cutting bee, when the pollen ball is complete, a single female lays an egg on top of the pollen ball and seals the brood cell. It differs from plain field-gathered pollen as honeybee secretions induce a fermentation process, where biochemical transformations break down the walls of flower pollen grains and render the nutrients more readily available. Bee pollen is harvested as food for humans and 'bee bread', due to the fermentation process, is much more potent than unadulterated flower pollen. It contains proteins, amino acids and a balanced composition of the mixture of carbohydrates (glucose, fructose), plant analogs of sex hormones, cell growth factors, minerals: calcium, potassium, magnesium, phosphorus, vitamin E, carotene, rutin, microelements such as copper, iron, sulfur, zinc, cobalt, gold, titanium and platinum. It has an impact on living organisms, including, for example, the mechanism of construction of major systems of newly-born bees. The weight of pollen-fed larvae increases by fifteen hundred times in three days! No one product in the world has such biological activity. There is never enough of this unique product – pollen produced by bees is in limited quantities. It cannot be artificially cultivated or synthesised, so ambrosia is considered the most rare and exclusive product of the hive.

Bee bread production and harvesting using special combs. (Alexander Komissar, Ukraine)

Apitoxin - bee venom

It is known that some of the most long-lived people are related to beekeeping. It is said that the secret to longevity lies in the fact that throughout the life they get many bee stings. Bee venom is one of the most powerful catalysts of physiological processes. Even the minimum of its concentration has a significant positive impact on the human body. The bee uses its venom only in self-defence in protecting the hive. By reducing the viscosity of the blood and reducing blood clots, bee venom is useful in the prevention of thrombotic complications of various locations (myocardial infarction, ischemic stroke, vascular thrombosis, and limb etc.). Apitoxin reduces the concentration of cholesterol in the blood, lowers blood pressure, prevents the development of arrhythmias, relieves angina, feeling of fatigue, low mood and insomnia. It is noted the analgesic effect of the venom (50 times stronger than Novocaine), it can be used for various pain syndromes. It adjusts the work of the human nervous system (lumbago, sciatica, low back pain, cerebral palsy, Parkinson's disease, multiple sclerosis, etc.), cures sores and varicose veins, chronic inflammation, sensorineural hearing loss, hyperthyroidism, psoriasis, and atopic dermatitis. Bee venom has an absorbing effect and the ability to smooth post-operative and burn scars. Apitherapy is used in asthma, eye diseases, infertility, menopause, menstrual irregularities, and impotence. Australian scientists from the Cancer Research Center at the Prince of Wales Hospital said that they could kill cancer cells using synthesized mellitin (the main component of bee venom). Current studies and investigations confirm that nanoparticles loaded with the bee venom mellitin destroy the AIDS-causing HI viruses without damaging other body cells.

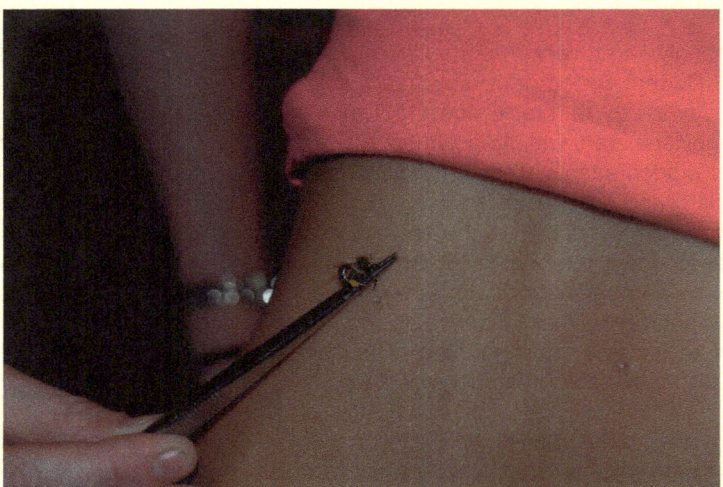

Bee stings are used for a wide range of health problems.

Ognevka - ОГНЕВКА

Wax moth species depend for their existence on the wax and detritus produced in the nests of honey bees and bumblebees. Like most members of the order of Lepidoptera, the larvae of many species are major pests in agriculture. The wax moth is the only known species of insects whose larvae feed exclusively on waste products of bees and have an amazing ability to digest wax. Wax moths are common wherever there are bees, with the exception of the high mountains and areas with harsh climates. The moth itself as an adult insect never eats; their lives are only sustained by the food they consumed as larvae. Many viruses and bacteria have a protective cover, resembling wax by its properties, that is very resistant to chemical attack. Only the digestive enzymes of wax moth larvae i.e. lipase (an enzyme that catalyzes the hydrolysis of fats, i.e. lipids) and tserraza (which promotes the softening and careful removal of surface skin cells and stimulates the skin renovation process) can break down the wax into its simple components. Microorganisms lose their protection and become unprotected. So was Ognevka, the consumption of wax moths, the first remedy against tuberculosis before the invention of antibiotics? The causative agent of tuberculosis, the tubercle bacillus, is protected by a wax like cover and can be destroyed by Ognevka. Nowadays it is well known that long-term use of Ognevka in food or as cosmetics has the best effect on human tissues and organs, regeneration and restoration, and has a rejuvenating effect. There are amazing things Ognevka can help with; it dissolves cholesterol plaque (cholesterol is a waxy product), strengthens blood vessels and the heart muscle. People have long known about the healing properties of wax moth larvae. They were eaten in the 17th century when it was recommended for consumptive patients, the elderly and childless women.

Interestingly, and very importantly regarding the environment, in 2017 the journal Current Biology published an article on the ability of the larvae of the large wax moth to break down polyethylene. Polyethylene and the similar polypropylene are synthetic polymers made from fossil petroleum products and are not biodegradable. They are mainly used in the production of plastic bags which is giving great concern worldwide through their haphazard disposal and accumulation, especially in the sea. The ability of the larvae to break down polyethylene is therefore an interesting approach to tackle this major environmental problem. The degradation rate, in tests, was found to be 0.23 mg cm^{-2} h^{-1}; this is significantly faster than the PE degradation rate of a recently discovered bacterium and the intestinal flora of the dry fruit moth *(Plodia interpunctella)*, which also has the ability to degrade PE.

Also of use in apitherapy is the Excrement of Wax Moth Larvae, in Russia called **"ЭКСКРЕМЕНТЫ ЛИЧИНОК ВОСКОВОЙ МОЛИ (ПЖВМ)"** in short PZHVM. The effect of the PZHVM tincture is more pronounced in comparison to the tincture of the wax moth larvae, the result being already noticeable in the first days (and often also hours) of it being administered. This is due to the fact that the tincture of PZHVM provides a lot of energy. PZHVM are collected when the wax moth larvae have eaten all the wax and it has already been digested several times by them (5-7 months). This ensures maximum saturation of the starting product with active ingredients and enzymes. Considering that the most active components of the wax moth larvae extract are its digestive enzymes, it can be safely said that the concentration of these enzymes in the wax moth excrements significantly exceeds the concentration of similar substances in which the larvae themselves have been processed.

Wax moth larvae and their detritus are valuable products for human health.

As we can see, we have lost a lot of knowledge in our culture due to useless wars, refuge, suppression, borders and the pressure of modern healthcare companies. It is time to remember the old knowledge and use it for ourselves rather than for it to be passed into the hands of pharmaceutical chemicals which will patent the products and sell them at exorbitant prices.

Besides being a beekeeper and alchemist Jörg is a practitioner of martial arts and a licensed trainer in traditional Japanese archery, Kyudo. His beekeeping philosophy is sustainable and treatment-free and he is researching traditional bee medicine, inspired by spagyric traditions. He studied Alchemy with "The Philosophers of Nature", founded by the French Alchemist Jean Dubuis who leads "Les Philosophes de la Nature" in France. Jörg is living with his family and his bees in a small village in the lower Rhine region not far from Düsseldorf.

Some Russian YouTube videos showing the process of making Podmor:
https://www.youtube.com/watch?v=Gjn84d94EDk
https://www.youtube.com/watch?v=YXAJnqS2CWc
https://www.youtube.com/watch?v=45vy1dbFAdU
https://www.youtube.com/watch?v=o7WWdkNFWfgI

R. Raff

HONEY
The never ending question - how much should I sell it for?

R Raff
From The Beekeepers Quarterly No 61, Spring 2000

In BKQ No 59, the editor wrote in his editorial on The Price of Honey. This is what he said:

Back in the UK perhaps we should use R. Raff's formula for fixing the price of honey, ie, relate it to the price of sugar. . . .. So we must start with a time when beekeepers thought that they were getting a fair price for honey, look at the price of sugar both then and now and make the necessary calculation. But where do we start from?

Well, let's go back say 45 years and have a look at a few other things beside sugar. For ease I'll give all the prices in decimal. It might confuse some if I gave all the prices in the old currency. Two pounds, twelve shilling and sixpence may not mean much. That was the price of a gross of jars in 1955.

In 1955 sugar was 7.3p a kilo and today it is 70p, roughly an increase of tenfold. Sugar has actually gone down in price over the past four years. In 1996 it was 79p a kilo and during these four years it has been as low as 69p.

In 1955 we were quite happy to get 15p for a one pound jar of blossom honey. At that time there was no VAT so we should make an addition for that and a price today of about £1.80/£2.00 wholesale might be about right. Some may be able to get more than that, a lot more indeed, but we must accept that a thing is only worth as much as we can get for it. The price of sections, cut comb and heather would have to be considered and adjustments made. With cut comb, for instance, although you lose nice drawn comb you save a lot of work.

Going through the items we use in our beekeeping I can see that over the years they, too, have gone up 10/12- fold, as a rule. We always have to remember the VAT though. In our craft we use jars, foundation and petrol. Since 1957: petrol has increased about 14-fold, jars have increased 10-fold (12-fold with VAT), foundation has increased about 9-fold, (10-fold with VAT).

There is, however, one item which has gone through the roof as far as price increase is concerned. For the sake of political correctness I'll call it the odd man out. Even calling it that could still get me into trouble. It could just as easily be the odd woman out. I am talking about CREOSOTE. I was 30 years of age before I lived in a house with electricity. Up to that time our home was served by coal gas as were most of the houses then. Creosote was a by-product of the coal gas industry and we got our creosote either from the gas works or the saw mill. It was the only wood preservative available at that time. It was dirt cheap and we even used it as a weedkiller splashing it around the hives to kill the weeds and grass. At the saw mill there was a massive tank with a door like a safe door. The tank would be filled with wood, fence posts for instance, and the creosote was pumped in at high pressure. That pressure would put the creosote clean through a 4" x 4" fence post. We went with our two or five-gallon can and the man would don rubber gloves, open a trap door in the floor and ladle out the hot creosote. At the gas works I never saw how he drew it off. In 1957 it cost 12 1/2p a gallon, 4 1/2 litres, roughly. We were sploshing the stuff around everywhere. Brood boxes were done both inside and outside as a measure against acarine. The hives were creosoted on a warm March day - and I mean the hives with bees in them. Very often the bees would be flying but they never seemed to mind. I've heard it said that the smell of creosote upsets them but I never found it so. I tell you, we were creosote crazy. Today, of course, you will be told that it is carcinogenic. Well, tell me that it is NOT carcinogenic now? If you believe and take heed of all that you read and hear, it is hardly worthwhile getting out of bed in the morning. In those days we were often strapped for cash and to stretch the creosote we would make a half-and-half mixture of creosote and old engine sump oil. I've often done it but I never liked it for the hives, especially the ones with the bees in them. This mixture took such a long time to dry. It was OK for sheds and fences. The addition of the oil used to give the end result a sort of skin that was more likely to shed water than preserve the wood. It did have the advantage though of being black, and it did absorb the heat of the sun in wintertime and that was a help to the bees. The winters then seemed to be much more severe than they are now. Global warming today I suppose.

So, here we are with creosote at 12 1/2p a gallon in 1955. I recently visited two big DIY stores and at one it was £4.94 a gallon and at the other it was £4.48. These were 4 litre containers and I've calculated it for a gallon. It is a sobering thought to think that the humble creosote today is so much dearer than petrol. In one store it was an increase of nearly 40-fold and in the other about 36-fold.

Petrol is not the best guide because the government plays fast and loose with it. In 1957 a gallon cost 25p whereas today it is £3.55, a 14-fold increase. To tell the truth I have considerable sympathy for the government when it jacks up taxes. Everybody is clamouring for more of this and that and the next thing. It all costs money and the money has to come from somewhere. Going back 45 years or so this country was awash with money. People were coming from all over the world to the UK to have major surgery performed free under the NHS. I am surprised the tax on whisky is not more than it is.

Like many beekeepers in the past, R Raff was a great believer in using creosote as a wood preserver. He gave me the tip that scorching the parts of posts that go into the ground helped to prevent them from rotting. It is important though to scorch the wood a few inches above ground level for it as that point most of the rotting first takes place.

In past years I have seen tables published annually with recommended prices for selling honey. They were, of course, only a guide but were a help. Beekeepers who sell honey should keep careful records of what all the items cost year by year. If for nothing else, these records will make interesting reading in later years. I have kept a daily journal for over fifty years and it is a wonderful thing for settling arguments about when things took place. It is good to create your own price structure and not bother about what others do.

I can remember in 1935 at work being sent out to buy fags. A twenty pack of Players cigarettes, with a book of matches thrown in, cost one shilling (5p). Today the increase is about 74-fold. In the matter of income it is interesting to see that in 1968 the state pension for a married couple was £7.30. Twenty-five years later, in 1993, it was £89.80, a 12-fold increase. It would seem that as long as you keep off fags and creosote we are not much worse off today than in previous years.

For some of our readers I can offer a crumb of comfort. In 1939, 61 years ago, a bottle of whisky was fourteen shillings and threepence; that is 71 1/2p in today's money. An average price today is about £11, although I have seen some selling at £8.99 and it was not in the paint stripper class either! That is an increase of 15-fold, but remember, that is 61 years ago. It's a bargain! I just mention that for interest, and maybe encouragement, but it does not give all those beekeepers who are down in the dumps about the price of honey, the price of equipment, the price of Bayvarol, and so on, licence to go to bed with a bottle of The Famous Grouse and a straw - or the teat of a baby's bottle.

PUZZLES

In Beekeeping Techniques, by Alexander Dean,1963 in the chapter on Statistical Techniques, he wrote:

The science of statistics is one which is looked upon with more than suspicion by many people. It is asserted by some that there are three types of lie: the white lie, the black lie and, statistics. Or again, it is said that figures can be made to prove anything.

Another common statement is that figures cannot lie: however, it is more accurate to go further and say that liars can figure. The following proof that 2 = I will illustrate this point and although we have some a *priori* evidence that this statement is absurd the flaw in the mathematical proof of a statement may not always be so obvious as in this particular instance.

Obviously, 2 cannot equal 1. Where is the fault in the equation below?

Proof that $2 = 1$.
1. Let $a = b$.
2. Multiply both sides of the equation by a: giving $a^2 = ab$.
3. Subtract b^2 from both sides giving $a^2 - b^2 = ab - b^2$.
4. Factorise: giving $(a-b)(a+b) = b(a-b)$.
5. Divide by $(a-b)$ giving $a+b = b$; since, however, $a = b$ we have $2 = 1$.

WORD SEARCH

This is taken from the first edition of The Beekeepers Annual. It contains the names of 37 authors of beekeeping books published before 1982. The answers can be found in the grid, the names being read vertically, horizontally or diagonally, with the letters in the right order or reversed.

L	L	U	N	D	W	E	R	G	I	T	T	E	P
U	L	M	A	C	E	V	R	E	L	T	U	B	D
B	O	A	N	A	W	O	C	N	N	U	M	R	D
B	B	N	S	H	E	R	Z	O	G	N	A	K	E
O	U	L	R	P	P	G	T	A	L	K	C	O	A
C	Z	E	E	I	M	L	Y	S	C	N	S	B	N
K	Z	Y	W	M	A	E	O	U	I	E	C	O	S
P	A	V	O	R	D	N	H	L	D	L	T	N	L
S	R	U	L	S	A	S	R	D	A	T	X	S	S
L	D	R	F	S	E	E	E	R	O	M	D	E	W
D	E	R	O	M	T	G	K	C	O	R	G	L	Y
N	E	F	L	E	A	P	D	R	S	G	R	S	T
E	C	R	A	N	E	N	S	O	I	S	T	E	P
R	L	M	A	N	L	E	Y	D	H	S	I	U	H

THE AUTHORS WORD SEARCH

Hidden in the grid are the names of 37 authors of beekeeping books which have been written since 1982.

```
S C Z E M I K S H A W K E R
I U S W E E N E Y S A L V O
M L E S H O W L E R D A Y B
S L E P H I P P S A X W D S
O U L Y W S T O R C H E G O
Q M E U K E A T K I N S O N
E K Y K B S I V A D Z N U M
H E H U T N T G E K Y E L A
E N X O G Z R I H U E S S N
A Y N T R E I U R T R R O G
F O O O G N C B T K M K N U
S N W O R B E G N I R A W M
X E R E G D A B Z C H I N G
N Y C R A M P Y A T E S K C
```

Books Published by Northern Bee Books in 2018

AN ECONOMICO-PHILOSOPHICAL DISCOURSE ON BEE CULTURE
VERNON, FRANK
ISBN: 9781912271276

THE ETHICS OF BEEKEEPING
WHITAKER, JOHN M
ISBN: 9781912271245

LA RUCHE DE LAYENS MODERNISEE
HURPIN, JEAN
ISBN: 9781912271269

BEEKEEPING IN VICTORIAN NOTTINGHAMSHIRE 1837 - 1901
CHING, STUART JOHN
ISBN: 9781912271238

THE SACRED BEE: IN ANCIENT TIMES AND FOLKLORE
RANSOME, HILDA M
ISBN: 9781912271214

YES! TOP BAR HIVES
SLADE, J. R.
ISBN: 9781912271221

NEW BEEKEEPING IN A LONG DEEP HIVE
DARTINGTON, ROBIN
ISBN: 9781912271191

BEEKEEPING STUDY NOTES FOR THE BBKA EXAMINATIONS: VOLUME 2
(MODULES 5, 6, 7 AND 8)
YATES, DAWN
ISBN: 9780905652726

CONSTRUCTION INFORMATION FOR DARTINGTON HIVES:
WITH FULL DETAILS FOR MAKING THE 'GARDEN' AND 'COUNTRY' MODELS OF THE DARTINGTON LONG DEEP HIVE
DARTINGTON, ROBIN
ISBN: 9781912271153

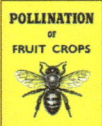
THE POLLINATION OF FRUIT CROPS
THE HORTICULTURAL EDUCATION ASSOCIATION,
ISBN: 9781912271184

THERE ARE QUEEN CELLS IN MY HIVE: - WHAT SHOULD I DO?
SHAW, WALLY
ISBN: 9781912271177

THE INTRODUCTION OF QUEEN BEES
ADAMS, BROTHER
ISBN: 9781912271160

THE GLASGOW BEEKEEPERS
GLASGOW AND DISTRICT BEEKEEPERS ASSOCIATION CENTENARY BOOK
EDITED BY TAYLOR HOOD
ISBN: 9781912271283

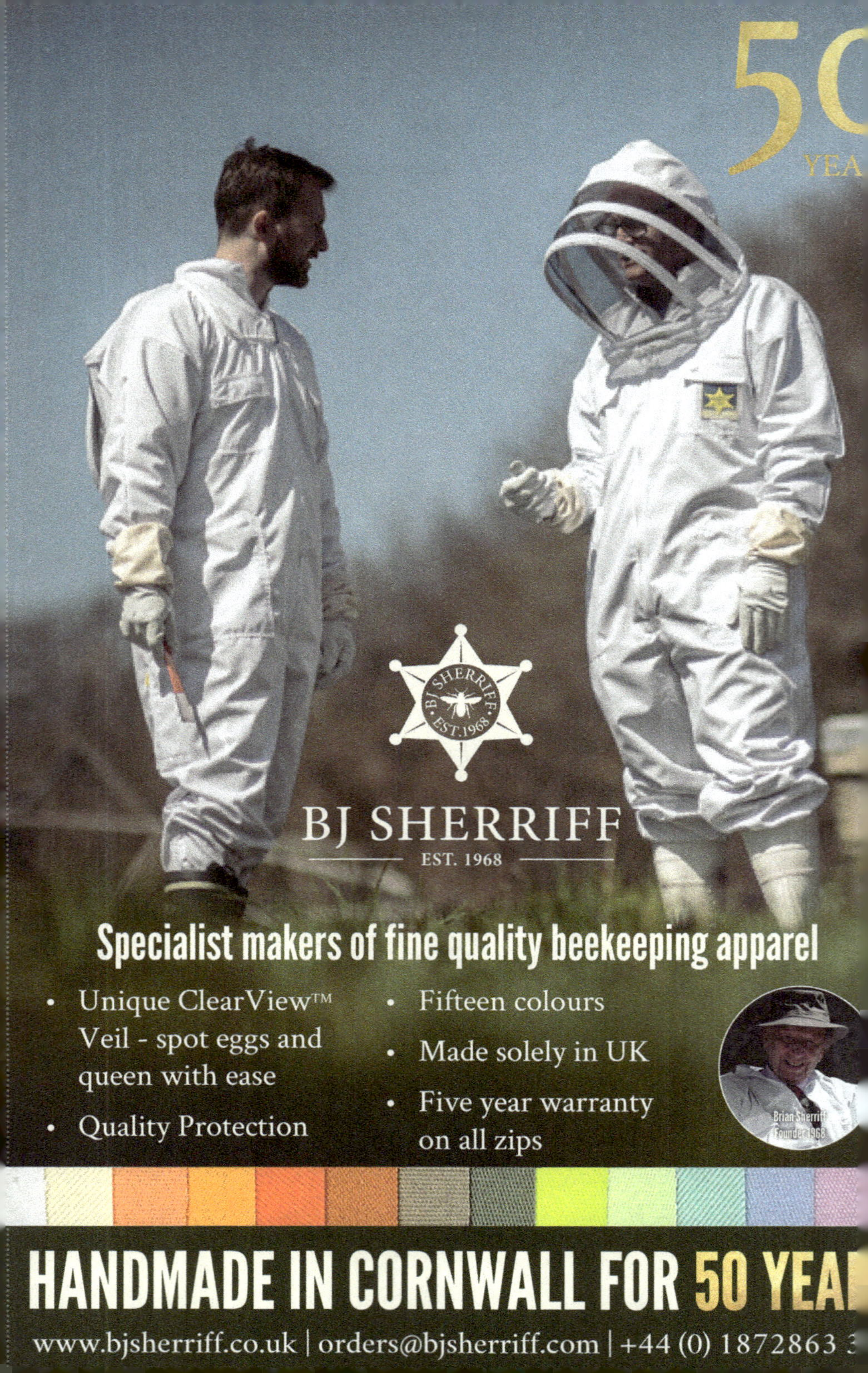

BJ SHERRIFF
EST. 1968

50 YEARS

Specialist makers of fine quality beekeeping apparel

- Unique ClearView™ Veil - spot eggs and queen with ease
- Quality Protection
- Fifteen colours
- Made solely in UK
- Five year warranty on all zips

Brian Sherriff
Founder 1968

HANDMADE IN CORNWALL FOR 50 YEARS

www.bjsherriff.co.uk | orders@bjsherriff.com | +44 (0) 1872863

19
DIARY & CALENDAR

- PART II -

***SR (SUNRISE am) SS (SUNSET pm) FOR LONDON UK.
WITH QUOTES FROM THE SHEPHERD'S CALENDAR AND
OTHER POEMS, BY JOHN CLARE (1793 - 1864).**

JANUARY

I love the snow, the crumpling snow
That hangs on everything,
It covers everything below
Like white dove's brooding wing,
A landscape to the aching sight,
A vast expanse of dazzling light.

DAY	JANUARY 2019 FORAGE	TEMP		WIND		CL'D	RAIN	1	2	3
		MIN	MAX	DIR	B.S			HIVE WEIGHT		
1										
2										
3										
4										
5										
6										
7										
8										
9										
10										
11										
12										
13										
14										
15										
16										
17										
18										
19										
20										
21										
22										
23										
24										
25										
26										
27										
28										
29										
30										
31										

JAN19

1,TU
NEW YEAR'S DAY

2,WE
SCOTLAND

3,TH

4,FR

5,SA
SR:08:06, SS:16:06

6,SU ○
EPIPHANY

7,MO
ORTHODOX CHRISTMAS

8,TU

9,WE

10,TH

11,FR

12,SA
SR:08:03, SS:16:15

13,SU

14,MO

15,TU

16,WE	**24,TH**
17,TH	**25,FR** BURN'S NIGHT
18,FR	**26,SA** SR:07:48, SS:16:38
19,SA SR:07:57, SS:16:26	27,SU
20,SU	**28,MO**
21,MO ● (WOLF) (SUPER MOON)	**29, TU**
22,TU	**30, WE**
23,WE	**31,TH**

FEBRUARY

Odd hive bees fancying winter o'er
And dreaming in their combs of spring
Creeps on the slab beside their door
And strokes its legs upon its wing
While wild ones half asleep are humming
Round snowdrop bells a feeble note

DAY	FEBRUARY 2019 FORAGE	TEMP		WIND		CL'D	RAIN	1	2	3
		MIN	MAX	DIR	B.S			HIVE WEIGHT		
1										
2										
3										
4										
5										
6										
7										
8										
9										
10										
11										
12										
13										
14										
15										
16										
17										
18										
19										
20										
21										
22										
23										
24										
25										
26										
27										
28										

FEB19

1,FR	**8,FR**
2,SA SR:07:38, SS:16:50 **CANDLEMAS**	**9,SA** SR:07:27, SS:17:04
3,SU	**10,SU**
4,MO ○	**11,MO** ST GOBNAIT, PATRON SAINT OF IRISH BEEKEEPERS
5,TU CHINESE NEW YEAR	**12,TU**
6,WE	**13,WE** ST MODMONOC, IRELAND
7,TH	**14,TH** ST VALENTINE
	15,FR

16,SA SR:07:14, SS:17:16	**24,SU**
17,SU	**25,MO**
18,MO	**26,TU**
19,TU ● (SNOW) (SUPER MOON)	**27,WE**
20,WE	**28,TH**
21,TH	
22,FR	
23,SA SR:06:59, SS:17:29 ST KHALAMPII, PATRON SAINT OF BULGARIAN BEEKEEPERS (HIVE-SHAPED PIES BAKED)	

MARCH

The Spring is come, and Spring flowers coming too,
The crocus, patty kay, the rich hearts' ease;
The polyanthus peeps with blebs of dew,
And daisy flowers; the buds swell on the trees;
While o'er the odd flowers swim grandfather bees

DAY	MARCH 2019 / FORAGE	TEMP MIN	TEMP MAX	WIND DIR	WIND B.S	CL'D	RAIN	1	2	3
								HIVE WEIGHT		
1										
2										
3										
4										
5										
6										
7										
8										
9										
10										
11										
12										
13										
14										
15										
16										
17										
18										
19										
20										
21										
22										
23										
24										
25										
26										
27										
28										
29										
30										
31										

MAR19

1,FR
ST. DAVID'S DAY (WALES)

2,SA
SR:06:45, SS:17:42

3,SU

4,MO

5,TU
SHROVE TUESDAY

6,WE ○
ASH WEDNESDAY

7,TH

8,FR
INTERNATIONAL WOMEN'S DAY

9,SA
SR:06:29, SS:17:54

10,SU

11,MO

12,TU

13,WE

14,TH

15,FR

16,SA SR:06:13, SS:18:06	**24,SU**
17,SU ST PATRICK'S DAY	**25,MO**
18,MO ST PATRICK'S DAY - OBSERVED N IRELAND	**26,TU**
19,TU	**27,WE**
20,WE SPRING EQUINOX	**28,TH**
21,TH ● (WORM) (SUPER MOON)	**29,FR**
22,FR	**30,SA** SR:05:41, SS:18:30 ST ALEXIUS DAY (UKRAINIAN BEEKEEPERS HANG ICONS OF THEIR PATRON SAINTS OF BEEKEEPING, ST SAVVATY AND ST ZOSIMA IN SHRINES AMONGST THEIR HIVES)
23,SA SR:05:57, SS:18:18 GREEK INDEPENDENCE DAY	**31,SU** BST BEGINS, MOTHERING SUNDAY

APRIL

Ploughmen go whistling to their toils
And yoke again the rested plough
And mingling o'er the mellow soils
Boys' shouts and whips are noising now

DAY	APRIL 2019 / FORAGE	TEMP		WIND		CL'D	RAIN	1	2	3
		MIN	MAX	DIR	B.S			HIVE WEIGHT		
1										
2										
3										
4										
5										
6										
7										
8										
9										
10										
11										
12										
13										
14										
15										
16										
17										
18										
19										
20										
21										
22										
23										
24										
25										
26										
27										
28										
29										
30										

APR19

1,MO

2,TU
SCOTLAND, SUPER MOON

3,WE

4,TH

5,FR ○ (PINK)

6,SA
SR:06:25, SS:19:42

7,SU

8,MO

9,TU

10,WE

11,TH

12,FR

13,SA
SR:06:10, SS:19:54

14,SU
PALM SUNDAY

15,MO

16,TU	**24,WE**
17,WE	**25,TH** ANZAC DAY
18,TH MAUNDY THURSDAY	**26,FR** ORTHODOX GOOD FRIDAY
19,FR ● GOOD FRIDAY	**27,SA** SR:05:40, SS:20:17 LAST DAY OF PASSOVER
20,SA SR:05:55, SS:20:05 HOLY SATURDAY, FIRST DAY OF PASSOVER	**28,SU** ORTHODOX EASTER
21,SU EASTER DAY	**29,MO** ORTHODOX EASTER MONDAY
22,MO EASTER MONDAY	**30,TU** ST ZOSIMA - 'GREET THE BEE ON ZOSIMA'S DAY AND THERE WILL BE HIVES AND WAX'.
23,TU ST. GEORGE'S DAY, SHAKESPEARE DAY	

MAY

To watch the bees that hang and swive
In clumps about each thronging hive
And flit and thicken in the light
While the old dame enjoys the sight
And raps the while their warming pans
A spell that superstition plans
To coax them in the garden bounds
As if they lov'd the tinkling sounds

DAY	MAY 2019 / FORAGE	TEMP		WIND		CL'D	RAIN	1	2	3
		MIN	MAX	DIR	B.S			HIVE WEIGHT		
1										
2										
3										
4										
5										
6										
7										
8										
9										
10										
11										
12										
13										
14										
15										
16										
17										
18										
19										
20										
21										
22										
23										
24										
25										
26										
27										
28										
29										
30										
31										

MAY 19

1,WE	**8,WE**
2,TH	**9,TH** LIBERATION DAY (GUERNSEY, JERSEY)
3,FR	**10,FR**
4,SA ○ SR:05:27, SS:20:29	**11,SA** SR:05:15 SS:20:40
5,SU MAY BANK HOLIDAY, RAMADAN BEGINS	**12,SU**
6,MO	**13,MO**
7,TU	**14,TU**
	15,WE

16,TH	**24,FR**
17,FR	25,SA SR:04:55, SS:21:00
18,SA ● (FLOWER) (BLUE MOON) SR:05:04, SS:20:51	26,SU
19,SU	**27,MO** SPRING BANK HOLIDAY
20,MO WORLD BEE DAY	**28,TU**
21,TU	**29,WE**
22,WE	**30,TH** ASCENSION DAY
23,TH	**31,FR**

JUNE

The clock-a-clay is creeping on the open bloom of May,
The merry bee is trampling the pinky threads all day,
And the chaffinch it is brooding on its grey mossy nest
In the whitethorn bush where I will lean upon my lover's breast;
I'll lean upon her breast and I'll whisper in her ear
That I cannot get a wink o'sleep for thinking of my dear;
I hunger at my meat and I daily fade away

DAY	JUNE 2019 / FORAGE	TEMP MIN	TEMP MAX	WIND DIR	WIND B.S	CL'D	RAIN	1	2	3
								HIVE WEIGHT		
1										
2										
3										
4										
5										
6										
7										
8										
9										
10										
11										
12										
13										
14										
15										
16										
17										
18										
19										
20										
21										
22										
23										
24										
25										
26										
27										
28										
29										
30										

JUN 19

1,SA SR:04:48, SS:21:09	**8,SA** SR:04:44, SS:21:16
2,SU	**9,SU** **PENTECOST**
3,MO ○	**10,MO** **WHIT MONDAY**
4,TU	**11,TU**
5,WE	**12,WE**
6,TH ST NICHOLAS	**13,TH**
7,FR	**14,FR**
	15,SA SR:04:42, SS:21:20

16,SU TRINITY SUNDAY, FATHER'S DAY	**24,MO**
17,MO ● (STRAWBERRY)	**25,TU**
18,TU	**26,WE**
19,WE	**27,TH**
20,TH	**28,FR**
21,FR SUMMER SOLSTICE	**29,SA** SR:04:45, SS:21:22
22,SA SR:04:42, SS:21:23 WINDRUSH DAY	**30,SU**
23,SU	

JULY

*Now love teaz'd maidens from their droning wheel
At the red hour of sunset sliving steals
From scolding dames to meet their swains again*

DAY	JULY 2019 / FORAGE	TEMP		WIND		CL'D	RAIN	1	2	3
		MIN	MAX	DIR	B.S			HIVE WEIGHT		
1										
2										
3										
4										
5										
6										
7										
8										
9										
10										
11										
12										
13										
14										
15										
16										
17										
18										
19										
20										
21										
22										
23										
24										
25										
26										
27										
28										
29										
30										
31										

JUL19

1,MO	**8,MO**
2,TU ○	**9,TU**
3,WE	**10,WE**
4,TH	**11,TH**
5,FR	**12,FR** BATTLE OF THE BOYNE (NORTHERN IRELAND)
6,SA SR:04:50, SS:21:20	13,SA SR:04:57, SS:21:14
7,SU	14,SU
	15,MO

16,TU ● (BUCK)	**24,WE**
17,WE	**25,TH**
18,TH	**26,FR**
19,FR	27,SA SR:05:16, SS:20:58
20,SA SR:05:06, SS:21:07	28,SU
21,SU	**29,MO**
22,MO	**30,TU**
23,TU	**31,WE**

AUGUST

Sweet solitude, what joy to be alone--
In wild, wood-shady dell to stay for hours.
T'would soften hearts if they were hard as stone
To see glad butterflies and smiling flowers.
Tis pleasant in these quiet lonely places,
Where not the voice of man our pleasure mars,
To see the little bees with coal black faces
Gathering sweets from little flowers like stars.

DAY	AUGUST 2019 / FORAGE	TEMP		WIND		CL'D	RAIN	1	2	3
		MIN	MAX	DIR	B.S			HIVE WEIGHT		
1										
2										
3										
4										
5										
6										
7										
8										
9										
10										
11										
12										
13										
14										
15										
16										
17										
18										
19										
20										
21										
22										
23										
24										
25										
26										
27										
28										
29										
30										
31										

AUG19

1,TH ○	**8,TH**
2,FR	**9,FR**
3,SA SR:05:26, SS:20:46	**10,SA** SR:05:37, SS:20:34
4,SU	**11,SU**
5,MO SUMMER BANK HOLIDAY (SCOTLAND)	**12,MO**
6,TU	**13,TU**
7,WE	**14,WE** THE SAVIOR OF THE HONEY FEAST (UKRAINE)
	15,TH ● (STURGEON) ASSUMPTION OF THE VIRGIN 'APOKIMISIS TIS PANAGIAS'

16,FR	24,SA SR:05:59, SS:20:05 ST BARTHOLOMEW (TRADITIONAL DAY FOR HARVESTING HONEY)
17,SA SR:05:48, SS:20:20	25,SU
18,SU	**26,MO** SUMMER BANK HOLIDAY (ENGLAND, WALES, NORTHERN IRELAND, GUERNSEY & JERSEY)
19,MO	27,TU
20,TU NATIONAL HONEY BEE DAY (USA)	28,WE
21,WE	29,TH
22,TH	30,FR
23,FR	31,SA SR:06:11, SS:19:50

SEPTEMBER

Black grows the southern sky, betokening rain,
And humming hive-bees homeward hurry bye:
They feel the change; so let us shun the grain
And take the broad road while our feet are dry.
Ay, there some dropples moistened on my face,
And pattered on my hat-otis coming nigh!

DAY	SEPTEMBER 2019 FORAGE	TEMP		WIND		CL'D	RAIN	1	2	3
		MIN	MAX	DIR	B.S			HIVE WEIGHT		
1										
2										
3										
4										
5										
6										
7										
8										
9										
10										
11										
12										
13										
14										
15										
16										
17										
18										
19										
20										
21										
22										
23										
24										
25										
26										
27										
28										
29										
30										

SEP19

1,SU	**8,SU**
2,MO	**9,MO**
3,TU	**10,TU**
4,WE	**11,WE**
5,TH	**12,TH**
6,FR	**13,FR**
7,SA SR:06:22, SS:19:34	**14,SA** ● (HARVEST) SR:06:33, SS:19:18
	15,SU

16,MO	**24,TU**
17,TU	25,WE
18,WE	**26,TH**
19,TH	**27,FR**
20,FR	**28,SA** SR:06:56, SS:18:46
21,SA SR:06:44, SS:19:02	29,SU
22,SU	**30,MO** ○
23,MO AUTUMN EQUINOX	

OCTOBER

Where the holly oak so tall
Far o'er tops the garden wall
That latest blooms for bees provide
Hived on stone benches close beside
The bees their teasing music hum
And threaten war to all that come
Save the old dame whose jealous care
Places a trapping bottle there
Filled with mock sweets in whose disguise
The honey loving hornet dies

DAY	OCTOBER 2019 FORAGE	TEMP		WIND		CL'D	RAIN	1	2	3
		MIN	MAX	DIR	B.S			HIVE WEIGHT		
1										
2										
3										
4										
5										
6										
7										
8										
9										
10										
11										
12										
13										
14										
15										
16										
17										
18										
19										
20										
21										
22										
23										
24										
25										
26										
27										
28										
29										
30										
31										

OCT19

1,TU	**8,TU**
2,WE	**9,WE** YOM KIPPUR
3,TH	**10,TH**
4,FR	**11,FR**
5,SA SR:07:07, SS:18:30	**12,SA** SR:07:19, SS:18:14
6,SU	**13,SU** ● (HUNTERS)
7,MO	**14,MO**
	15,TU

16,WE	**24,TH**
17,TH	**25,FR**
18,FR	26,SA SR:07:43, SS:17:45
19,SA SR:07:31, SS:17:59	27,SU OCT BST ENDS
20,SU	**28,MO** ○ 'OHI DAY' GREECE ('NO TO MUSSOLINI) THANKSGIVING DAY, USA
21,MO	**29, TU**
22,TU	**30, WE**
23,WE	**31,TH**

NOVEMBER

And gobbling turkey cock wi noises vile
Dropping his snout as flaming as a cloak
Loose as a red rag o'er his beak the while
Urging the dame to turn her round and smile

DAY	NOVEMBER 2019 FORAGE	TEMP		WIND		CL'D	RAIN	1	2	3
		MIN	MAX	DIR	B.S			HIVE WEIGHT		
1										
2										
3										
4										
5										
6										
7										
8										
9										
10										
11										
12										
13										
14										
15										
16										
17										
18										
19										
20										
21										
22										
23										
24										
25										
26										
27										
28										
29										
30										

NOV19

1,FR
ALL SAINTS DAY

2,SA
SR:06:56, SS:16:32
ALL SOULS DAY

3,SU

4,MO

5,TU
GUY FAWKES DAY

6,WE

7,TH

8,FR

9,SA
SR:07:08, SS:16:20

10,SU
REMEMBRANCE SUNDAY

11,MO

12,TU ● (BEAVER)

13,WE

14,TH

15,FR

16,SA SR:07:20, SS:16:10	**24,SU**
17,SU	**25,MO**
18,MO	**26,TU** ○
19,TU	**27,WE**
20,WE	**28,TH**
21,TH	**29,FR**
22,FR	**30,SA** SR:07:43, SS:15:55 ST ANDREW'S DAY
23,SA SR:07:32, SS:16:01	

DECEMBER

Christmas:
Each house is swept the day before
And windows stuck wi evergreens
The snow is besomed from the door
And comfort crowns the cottage scenes
Gilt holly wi its thorny pricks
And yew and box wi berrys small
These deck the unused candlesticks
And pictures hanging by the wall

DAY	DECEMBER 2019 FORAGE	TEMP		WIND		CL'D	RAIN	1	2	3
		MIN	MAX	DIR	B.S			HIVE WEIGHT		
1										
2										
3										
4										
5										
6										
7										
8										
9										
10										
11										
12										
13										
14										
15										
16										
17										
18										
19										
20										
21										
22										
23										
24										
25										
26										
27										
28										
29										
30										
31										

DEC19

1,SU **ADVENT SUNDAY**	**8,SU** FEAST OF THE IMMACULATE CONCEPTION
2,MO ST ANDREW'S DAY OBSERVED (SCOTLAND)	**9,MO**
3,TU	**10,TU**
4,WE	**11,WE**
5,TH	**12,TH** ● (COLD)
6,FR **ST NICHOLAS**	**13,FR**
7,SA SR:07:51, SS:15:53 ST AMBROSE (PATRON SAINT OF BEEKEEPERS)	**14,SA** SR:07:59, SS:15:51
	15,SU

16,MO	**24,TU**
17,TU	**25,WE** DEC CHRISTMAS DAY
18,WE	**26,TH** ○ DEC BOXING DAY
19,TH	**27,FR**
20,FR	28,SA SR:08:07, SS:15:57
21,SA SR:08:04, SS:15:52	29,SU
22,SU WINTER SOLSTICE	**30,MO**
23,MO	**31, TU** NEW YEAR'S EVE

Hive/Q NO.	Year Q Raised	Frames of Brood Autumn 2018	Combs Covered	Honey Stored-Sugar fed Kg	Combs Covered Spring 2019	Frames of Brood Spring 2019	Spring Feeding Kg	Queens Reared	Nuclei
1									
2									
3									
4									
5									
6									
7									
8									
9									
10									
11									
12									
13									
14									
15									
16									
17									
18									
19									
20									
21									
22									
23									
24									

HONEYBEE COLONIES

1									
2									
3									
4									
5									
6									
7									
8									
9									
10									
11									
12									
13									
14									
15									
16									
17									
18									
19									
20									
21									
22									
23									
24									

BEEEKEEPING RECORDS

Number	items	Est. Value £	P
	Stocks of Bees		
	Empty Hives		
	Combs - Deep - Shallow		
	Frames		
	Foundations		
	Honey Extractor		
	Honey Tanks		
	Other items		
	Honey Jars		
	Honey		

JANUARY 2020

S	M	T	W	T	F	S
			1	2	3	4
5	6	7	8	9	10	11
12	13	14	15	16	17	18
19	20	21	22	23	24	25
26	27	28	29	30	31	

FEBRUARY 2020

S	M	T	W	T	F	S
						1
2	3	4	5	6	7	8
9	10	11	12	13	14	15
16	17	18	19	20	21	22
23	24	25	26	27	28	29

MARCH 2020

S	M	T	W	T	F	S
1	2	3	4	5	6	7
8	9	10	11	12	13	14
15	16	17	18	19	20	21
22	23	24	25	26	27	28
29	30	31				

APRIL 2020

S	M	T	W	T	F	S
			1	2	3	4
5	6	7	8	9	10	11
12	13	14	15	16	17	18
19	20	21	22	23	24	25
26	27	28	29	30		

MAY 2020

S	M	T	W	T	F	S
					1	2
3	4	5	6	7	8	9
10	11	12	13	14	15	16
17	18	19	20	21	22	23
24	25	26	27	28	29	30
31						

JUNE 2020

S	M	T	W	T	F	S
	1	2	3	4	5	6
7	8	9	10	11	12	13
14	15	16	17	18	19	20
21	22	23	24	25	26	27
28	29	30				

JULY 2020

S	M	T	W	T	F	S
			1	2	3	4
5	6	7	8	9	10	11
12	13	14	15	16	17	18
19	20	21	22	23	24	25
26	27	28	29	30	31	

AUGUST 2020

S	M	T	W	T	F	S
						1
2	3	4	5	6	7	8
9	10	11	12	13	14	15
16	17	18	19	20	21	22
23	24	25	26	27	28	29
30	31					

SEPTEMBER 2020

S	M	T	W	T	F	S
		1	2	3	4	5
6	7	8	9	10	11	12
13	14	15	16	17	18	19
20	21	22	23	24	25	26
27	28	29	30			

OCTOBER 2020

S	M	T	W	T	F	S
				1	2	3
4	5	6	7	8	9	10
11	12	13	14	15	16	17
18	19	20	21	22	23	24
25	26	27	28	29	30	31

NOVEMBER 2020

S	M	T	W	T	F	S
1	2	3	4	5	6	7
8	9	10	11	12	13	14
15	16	17	18	19	20	21
22	23	24	25	26	27	28
29	30					

DECEMBER 2020

S	M	T	W	T	F	S
		1	2	3	4	5
6	7	8	9	10	11	12
13	14	15	16	17	18	19
20	21	22	23	24	25	26
27	28	29	30	31		

DIRECTORY 19
DIRECTORY, Associations and Services

Every effort is made to keep entries up to date but the publishers cannot be held responsible for errors or omissions.
Associations and all other groups listed have been requested (August 2018) to supply updated entries.
Readers who are aware of inaccuracies are asked to send updates to jerry@northernbeebooks.co.uk

BEEKEEPING ASSOCIATIONS
Bee Educated e-learning for Beekeepers

http://www.beeeducated.co.uk/
ian@BeeEducated.co.uk

EDU
e-Learning for Beekeepers
A module website set-up specifically for beekeeping tutors and their students
http://www.beeeducated.co.uk/
mail: st@zbee.com

BDI
Bee Disease Insurance Ltd
http://www.beediseasesinsurance.co.uk/home
donald.robertson-adams@beediseasesinsurance.co.uk

BEES
Beekeeping Editors' Exchange Scheme
Helping editors help themselves
editors-owner@ebees.org.uk

BA
Bees Abroad
Relieving poverty through beekeeping
http://beesabroad.org.uk/
info@beesabroad.org.uk
www.facebook.com/beesabroad/

BFA
Bee Farmers Association
The voice of professional beekeeping
http://beefarmers.co.uk/
gensec@beefarmers.co.uk
Publication: Bee Farmer Journal
www.facebook.com/beefarmersassociation

B *for* D
Bees for Development
The specialist international beekeeping organisation
http://www.beesfordevelopment.org/
info@beesfordevelopment.org
www.facebook.com/beesfordevelopment/
Publication: Bees for Development Journal

BBKA
British Beekeepers Association
https://www.bbka.org.uk/
www.facebook.com/groups/BBKA.info/
Publication: BBKA News

BIBBA
British Improvement and Bee Breeders' Association
For the conservation, restoration, study, selection and improvement of native or near native honeybees of Britain and Ireland
https://bibba.com/
membership@bibba.com
Publication: Bee Improvement Magazine
www.facebook.com/beeimprovement

CABK
Central Association of Beekeepers
https://www.cabk.org.uk/
pamhunter@burnthouse.org.uk
Publications: Selected lectures

CBDBBRT
C.B. Dennis British Beekeepers Trust
https://sites.google.com/site/cbdennistrust/
cbdennisbeetrust@gmail.com

CONBA
Council of National Beekeeping Association of the UK
Its purpose is to represent the interests of beekeepers with local, national and international authorities
http://www.conba.org.uk/
mail: philmcanespie@btinternet.com

DARG
Devon Apicultural Research Group
Combining practical beekeeping with leading edge apicultural and environmental science
http://dargbees.org.uk/

ECT
The Eva Crane Trust
To advance the understanding of bees and beekeeping
https://www.evacranetrust.org/
mail@evacranetrust.org

FIBKA
Federation of Irish Beekeepers' Association
https://irishbeekeeping.ie/
secretary@irishbeekeeping.ie
www.facebook.com/fibka/
Publication: An Beachaire

IBRA
International Bee Research Association
Promotes the value of bees by providing information on bee science and beekeeping worldwide
http://www.ibrabee.org.uk/
mail@ibra.org.uk
www.facebook.com/IBRAssociation/
Publications: Bee World, Journal of Apicultural Research

INIB
Institute of Northern Ireland Beekeepers
http://www.inibeekeepers.com/
membershipsecretary@inibeekeepers.com

LASI
Laboratory of Apiculture and Social Insects
The applied research is aimed at helping the honey bee and beekeepers, whilst the basic research studies how insect societies function
http://www.sussex.ac.uk/lasi/
www.facebook.com/

NBKT
Natural Beekeeping Trust
We encourage attention to the real nature of bees, their nesting preferences, their forage needs and their all-encompassing purpose.
https://www.naturalbeekeepingtrust.org/
www.facebook.com/naturalbeekeepingtrust/

NDB
National Diploma in Beekeeping
The NDB exists to meet a need for a beekeeping qualification above the level of the Certificates awarded by the United Kingdom National Beekeeping Associations
https://national-diploma-beekeeping.org/

NHS
National Honey Show
Promoting the highest quality honey and wax products with international classes, lecture convention, workshops and beekeeping equipment trade show
http://www.honeyshow.co.uk/

NIHBS
Native Irish Honey Bee Society
To support the various strains of Native Irish Honey Bee (Apis mellifera mellifera) throughout the country. It is a cross border organisation and is open to all. It consists of members and representatives from all corners of the island of Ireland
http://nihbs.org/
secretary@nihbs.org
www.facebook.com/native-irish-honey

SBA
Scottish Beekeepers' Association
The organisation's purposes are to support honey bees and beekeepers, to improve the standard of beekeeping, and to promote honey bee products in Scotland
https://scottishbeekeepers.org.uk/
secretary@scottishbeekeepers.org.uk
Publication: The Scottish Beekeeper

UBKA
Ulster Beekeepers' Association
https://www.ubka.org/
ubkasecretary@gmail.com
www.facebook.com/Ulsterbees/

WBKA
Welsh Beekeepers' Association
We represent Welsh beekeepers nationally within Wales and the UK and internationaly.
http://www.wbka.com/
secretary@wbka.com
Publication: Welsh Beekeeper

HONEY BEE HEALTH
England and Wales
NBU
National Bee Unit
http://www.nationalbeeunit.com/
nbu@apha.gsi.gov.uk

Northern Ireland
DARD
Department of Agriculture Northern Ireland
https://www.daera-ni.gov.uk/

Scotland
SG-AFRC
The Scottish Government Rural Payments and Inspections Directorate
bees_mailbox@gov.scot

FACEBOOK BEEKEEPING GROUPS

Beekeeping for Beginners
Beekeeping Basics
Friendly Beekeepers
Beekeeping Hacks
Women in Beekeeping
The Beekeepers Bulletin
Poly hive beekeeper
Commercial Beekeepers
HAPPY BEEKEEPERS
UK Beekeepers
Cold Climate Beekeeping
Stewart's Beekeeping Basics
Newbie Beekeeping
Warre Beekeeping
Scientific Beekeeping
Backyard Beekeeping
British Beekeepers
Scottish Beekeepers
Beekeeping
Sideliner Beekeepers
Loui's Mountain Beekeeping
Beekeepers of Ireland
Beekeeping for beginners to experienced
London Beekeepers Association
Beekeeping Tools, Supplies & Hardware
Beekeeping with the warré hive
Beekeeping Classifieds
Florida Beekeepers
Northern Beekeeping
Irish Beekeepers' Association
Beekeeping Wood Shop
Buy/Sell Beekeeping Equipment (Serious Buyers)
Beekeeper Builders Corner
Top Bar Hive Beekeepers
Beekeeping Victoria (Australia)
Beekeeping for Beginners UK
Beekeeping Top Tips UK
Beekeepers buying or selling UK
Beekeeping in France
"OTS" Beekeeping
Lincolnshire Beekeeping Association- Sleaford District

- Backyard beekeeping NZ
- Beekeeping Science
- Too Big to Quit Now - Sideliner Beekeepers
- Beekeeping
- Backyard Beekeeping Australia
- Beekeeping for all
- Beekeeping Questions UK
- Treatment-free beekeepers
- Beekeepers Trading Post
- Better Beekeeping
- Beekeeping Techniques
- Beekeeping Apimarket UK
- beekeeping Warm and cold Large and small
- Beekeeping Classifieds Victoria Australia
- Jersey Beekeepers' Association
- Black bee beekeepers
- Beekeeping Equipment For Sale Australia
- Northamptonshire Treatment Free BeeKeeping
- Beekeeping
- Modern BeeKeepers
- Beekeepers Group :D
- Beekeeping Petra
- amateur backyard beekeeping
- Beekeeping Supplies eBay/Amazon
- World Beekeepers Collective
- Beekeeping for all
- Urban Beekeeper
- NWA Beekeeping Alliance
- Beekeeping Basics
- Successful Beekeeping
- Beekeeping Questions
- Irish Beekeeping
- EN-Beekeepers
- Ohio beekeepers
- Beekeeping Qld Australia
- Beekeepers World Club
- Natural and Balanced Beekeeping DIscussion Group
- Buckfast Beekeepers Group
- Beekeeping
- Commercial Beekeepers Trucks And Equipment For Sale Trade Etc
- Louth Beekeepers
- Preservation Beekeeping Community Page
- Beekeepers With Power!!

Apicultrices - Women in Beekeeping
Beekeepers Kenya
UK Beekeepers: Buy & Sell
BeekeepersOfCT
Beekeeping Photography
Beekeepers of Ireland BUY/SELL/SWAP/FREE
COBA Central Oklahoma Beekeepers Association

Suppliers of Beekeeping Equipment, etc.
www.facebook.com/E.H.Thorne/
www.facebook.com/BJSherriff/
www.facebook.com/BeeEquipmentLtd/
www.facebook.com/cornishhoney.co.uk/
www.facebook.com/quincehoneyfarm/
www.facebook.com/NaturalApiary/
www.facebook.com/BBwear/
www.facebook.com/paynesbeefarm/
www.facebook.com/maisemoreapiaries/
www.facebook.com/nationalbeesupplies/
www.facebook.com/BSHoneyBees/
www.facebook.com/Beckysbees/

Books:
Northern Bee Books
01326290265
www.northernbeebooks.co.uk
t: https://twitter.com/nthnbeebooks
f: https://www.facebook.com/beekeepingbooks
i: https://www.instagram.com/northernbeebooks

Collector's Corner

Beekeepers who collect postage stamps related to the craft, will be pleased to learn that ELTA, the Greek Post Office have just issued new stamps featuring Honey bees, *Apis mellifera*.

ANSWERS TO WORD SEARCH 2

ASTON	HEAF	SEELEY
ATKINSON	HORN	SHOWLER
BADGER	KIRK	SIMS
BROWN	KRITSKY	STORCH
CHING	LAWES	SWEENEY
CRAMP	MANGUM	TURNBULL
CULLUMKENYON	MIKSHA	WARING
DARTINGTON	MUNZDAVIS	WEIGHTMAN
DEBRUYN	OWEN	WHITTAKER
GOULSON	PHIPPS	YATES
GREGORY	RICE	
HAWKER	ROBSON	

www.ingramcontent.com/pod-product-compliance
Lightning Source LLC
Chambersburg PA
CBHW070909160426
43193CB00011B/1412